Risk Communication
and Miscommunication

Risk Communication and Miscommunication

CASE STUDIES IN SCIENCE, TECHNOLOGY, ENGINEERING, GOVERNMENT, AND COMMUNITY ORGANIZATIONS

Carolyn R. Boiarsky

UNIVERSITY PRESS OF COLORADO
Boulder

Published by University Press of Colorado
5589 Arapahoe Avenue, Suite 206C
Boulder, Colorado 80303

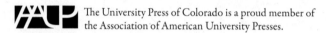 The University Press of Colorado is a proud member of
the Association of American University Presses.

The University Press of Colorado is a cooperative publishing enterprise supported, in part,
by Adams State University, Colorado State University, Fort Lewis College, Metropolitan
State University of Denver, Regis University, University of Colorado, University of
Northern Colorado, Utah State University, and Western State Colorado University.

∞ This paper meets the requirements of the ANSI/NISO Z39.48-1992 (Permanence of
Paper).

ISBN: 978-1-60732-466-9 (pbk)
ISBN: 978-1-60732-467-6 (ebook)

Library of Congress Cataloging-in-Publication Data

Names: Boiarsky, Carolyn R., author.
Title: Risk communication and miscommunication : case studies in science,
    technology, engineering, government and community organizations / by
    Carolyn Boiarsky, Ph.D., Purdue University—Calumet, Hammond, IN.
Description: Boulder : University Press of Colorado, 2016. | Includes
    bibliographical references.
Identifiers: LCCN 2015029798 | ISBN 9781607324669 (pbk.) | ISBN 9781607324676
    (ebook)
Subjects: LCSH: Risk communication. | Miscommunication.
Classification: LCC T10.68 .B65 2016 | DDC 658.4/5—dc23
LC record available at http://lccn.loc.gov/2015029798

IN MEMORIAM

To my husband Clement S. Stacy.

To the many men and women in the environmental sciences who endeavor to carry out their work responsibly to protect our planet. Their jobs are wrought with conflict that they must navigate to the best of their ability.

Contents

Acknowledgments

I received help from many people who took the time to provide me with copies of the original documents that are in this book and who willingly tried to answer my questions so that the record would be correct and complete.

How to Read This Book

Readers of this book may not be familiar with some of the crisis situations and environmental and technical disasters discussed. For background information and additional details, a list of websites is included at the conclusion of the book. Sites are listed by chapter.

Risk Communication
and Miscommunication

Introduction

Frequently, environmental communication occurs in a context that is controversial at the very least and hostile at its most contentious. The worst environmental disasters in the past fifty years—the Daiichi, Japan, and Three Mile Island, Pennsylvania, nuclear accidents; the Massey, West Virginia, coal mine explosion; and the British Petroleum (BP) Gulf oil spill—have all been at least partially related to written communication that has failed to take into consideration a "non-safety industrial culture" in which the "normalization of deviance" has become an acceptable practice and in which negative news is frowned upon, prompting a "week's vacation at the beach" (shorthand for a week's suspension). The president's report on the BP Gulf oil spill (2011) concludes, "Most, if not all, of the failures of Macondo [oil well] can be traced back to underlying failures of management and *communication* [emphasis added] . . . better communication within and between BP and its contractors . . . would have prevented the Macondo incident."[1] This statement echoes comments by the Columbia [shuttle] Accident Investigation Board. "Cultural traits and organizational practices detrimental to safety were allowed to develop, including: . . .

DOI: 10.5876/9781607324676.c000

organizational barriers that prevented effective communication of critical safety information."[2]

Because local communities, situated in potentially hazardous environments, such as those located in flood and hurricane zones or near industries involved with toxic wastes or fracking, are often split between factions of the community that fear potential environmental problems and those that perceive economic prosperity as a priority, effective communication between sides is often difficult. Such was the case in northwest Illinois when a California company wanted to locate a megafarm where local citizens were concerned about the potential leaching of toxic wastes into the water supply.

Whether or not readers successfully process a text so they can make appropriate decisions in an at-risk environment or in a crisis situation depends on whether the writers have taken into consideration the context in which the documents will be read and the pattern in which readers will read the messages. In the case of such tragic events as the shuttle, or flood and mine disasters, the writers failed to consider the readers' contexts and reading patterns. As a result, the readers failed to understand and follow through on the memoranda and reports that they received, because they either misunderstood or completely failed to understand a message. Readers might have recognized the problems and the potential for disaster if the messages had been clearer, the material had been organized in another pattern, the tone had been different, the focus had been on another aspect, or additional information had been included.

Background

Throughout history, environmental problems have resulted from industrialization as well as natural events. Often those designing, constructing, or overseeing commercial and industrial projects have either failed to recognize the environmental problems that the projects could cause or have refused to be concerned with them.

According to Stephen Ambrose in his book *Nothing Like It in the World: The Men Who Built the Transcontinental Railroad 1863–1869*, Americans in general and in industry specifically have always wanted their projects built "fast" at the cost of "well." "In America it was common practice to get the [rail]road open for traffic in the cheapest manner possible, and in the least possible time. The attitude was 'It can be fixed up and improved later.'"[3] Collis Huntington, one of the "Big Four" who was responsible for obtaining the financing for the Central Pacific Railroad, expressed this sentiment in 1868 in a message to Charlie Crocker, one of the other "Big Four" who was charged with overseeing the construction of the [rail]road[bed], "I would build the cheapest road that I could and have it accepted by the [government] Comm[ission] so it moves ahead fast."[4]

Building fast was the tendency followed by those who constructed the Hoover Dam. Luckily, the dam never gave way before the engineers could return to fix it once it was completed. According to Michael Hiltzik in his book *Colossus: Hoover Dam and the Making of the American Century*, Reclamation officials could not predict the "effect this unprecedented man-made load might have on the earth's malleable crust. One could only hope that the consequences would not be catastrophic."[5] The dam was completed in 1936; within eighteen months, water began seeping through its foundation walls into the galleries, causing the reclamation technical team to recommend redoing the grouting. The repairs took nine years to complete, nearly twice as long as it had taken to build the dam itself.[6] The twentieth century continued this modus operandi. The Accident Investigation Board of the *Columbia* shuttle disintegration suggested that during the 1990s, NASA's motto became "faster, better, cheaper."[7]

More recently, the BP Gulf oil spill of 2010 was partially caused by management's determination to get the rig built "fast." The national commission's report to the president, "Deep Water: The Gulf Oil Disaster and the Future of Offshore Drilling," indicates that both the engineers and the managers were in a rush to complete the job so that BP could start withdrawing the oil and thereby begin to draw earnings, and the engineers could move off what they perceived as a "nightmare well" (see Figure 4.9). In one case when told that a decision had been made to use fewer casings than usually required to hold a liner that would create a barrier to the flow of gas into inappropriate areas, an engineer commented in an e-mail, "Who cares, it's done, end of story, will probably be fine" (Figure 0.1).

From: Cocales, Brett W
Sent: Friday, April 16,2010 4:15 PM
To: Morel, Brian P
Subject: RE: Macondo STK geodetic
Even if the hole is perfectly straight, a straight piece of pipe even in
tension will not seek the perfect center of the hole unless it has
something to centralize it.
But, who cares, it's done, end of story, will probably be fine and we'll
get a good cement job. I would rather have to squeeze than get stuck
above the WH. So Guide is right on the risk/reward equation.

Best Regards

FIGURE 0.1. *BP/*Horizon *oil rig e-mail*

According to the President's Report, "Many of the decisions that
BP, Halliburton, and Transocean made that increased the risk of the
Macondo blowout clearly saved those companies significant time."[8]

It is because of this philosophy that writers in these companies often
fail to communicate the full extent of a problem, if they communicate
about it at all.

The Chicago flood memorandum and the Macondo e-mails were
written by writers who failed to take into consideration what their
readers knew, how their readers would read, and the environment in
which their readers would read. They wrote writer-based rather than
reader-based texts.[9] Although the documents were in different genres
and media, the problems of miscommunication were similar. Whether
documents are written as traditional letters and memos or whether they
are in the form of e-mails, text messages, or slideware (PowerPoint)
scripts, the need to write reader-based documents remains the same.
Some of the documents in this book were written before electronic
media became ubiquitous. These forms remain current even if the
mode of transmission has changed.

The need to write texts that consider the readers and the context in
which a document will be read is especially important in the environmental sciences. The day-to-day business of an organization depends
on the effective transmission of information internally as well as exter-
nally to clients, customers, and the surrounding community. While
much of the communication that takes place in the environmental

sciences is related to routine business matters, some, like that related to offshore drilling, fracking, and the manufacture of toxic wastes, has far-reaching effects, as we have seen. A company may decide to expand its environmental efforts because of a well-written report, and a state legislature may pass a bill to clean up local pollution in response to a persuasive proposal. But multimillion-dollar lawsuits for an injured employee or consumer or even an entire community may result from inadequately written procedures, as in the Three Mile Island nuclear accident:

> We [President's Commission investigating accident] found that the specific operating procedures, which were applicable to this accident, are at least very confusing and could be read in such a way as to lead the operators to take the incorrect actions they did . . . The lessons from previous accidents did not result in new, clear instructions being passed on to the operators. Both points are illustrated in the following case history. A senior engineer of the Babcock & Wilcox Company (suppliers of the nuclear steam system) noted in an earlier accident, bearing strong similarities to the one at Three Mile Island, that operators had mistakenly turned off the emergency cooling system. He pointed out that we were lucky that the circumstances under which this error was committed did not lead to a serious accident and warned that under other circumstances (like those that would later exist at Three Mile Island), a very serious accident could result. He urged, in the strongest terms, that clear instructions be passed on to the operators.[10]

Poorly organized documents, such as the PowerPoint presentation involved in the *Columbia* shuttle disaster, can also have disastrous consequences (see chapter 4).

Every document is written in a context that involves three aspects: (1) the reader; (2) the purpose, which may be the same for both the reader and the writer or may differ; and (3) the situation, which involves political, economic, social, and psychological factors. For example, a writer who sends a memorandum that requests members of a team to meet on a specific day and time may be able to persuade everyone on the team to attend if the members are willing to cooperate with the writer (the social and/or political context), if they believe that the meeting is worthwhile for the good of the company (the economic context), and if they are not threatened by the meeting (the psychological context). When these aspects are negative, as they were in the case of the *Columbia* space shuttle accident, communication breaks down.[11]

The communication problems discussed are representative of the problems that can be found in the messages involved with many major disasters during the past fifty years, regardless of whether they are communicated as traditional letters and memoranda or are transmitted electronically as e-mail and text messages. Communication that occurs during crisis situations involves helping those affected deal with the situation, but risk communication prior to such incidents—if it is effective—may be able to prevent such situations from occurring in the first place. This book is concerned with both kinds of communication.

Chapter 1 is concerned with the basic concepts underlying all communication: written or oral, read in hard copy or on electronic media. The chapter uses documents from the 1992 Chicago flood to exemplify how readers, their purposes, their reading patterns and styles, and the contexts in which they read a message affect their response to a communication. The second half of the chapter provides recommendations on ways to write reader-based documents.

Chapter 2 concentrates on strategies related to content, organization, and style that provide effective communication. Messages that were sent to residents in communities affected by the 2011 Mississippi flood are used to exemplify effective communication.

Chapter 3 is specifically concerned with strategies for persuading readers to agree with a writer's opinion. Web documents from opposing sides of the clean coal debate are used to exemplify these strategies.

Chapter 4 looks at e-mail and text transactions. It examines writer and reader interactions as threads in a message chain. Messages exchanged during the *Columbia* shuttle accident and prior to the BP/*Horizon* Gulf oil rig explosion are used to demonstrate effective and ineffective communication.

Chapter 5 moves into oral presentations with an examination of PowerPoint and slideware in general. Using slides from the army,

Challenger shuttle accident, and the Enbridge Pipeline, the chapter discusses the problems of using slideware and provides recommendations for designing and writing effective presentations.

Each chapter concludes with a summary of strategies for effective communication.

With the need to communicate complex and often controversial issues across vast geographic and cultural spaces with an ever-expanding array of electronic media, clear communication to a varied audience, including members of a community as well as professionals in a field, becomes increasingly important. As the weather brings more formidable storms with potential flooding, as the new shale oil lands open up increased potential for pipeline leaks and train derailments, and as coal remains a cheap and available source of power as well as pollution, it becomes incumbent upon those who work in these fields to understand the effects of their messages on their readers and to acquire the strategies necessary to communicate their data, ideas, and procedures effectively. It is hoped that the cases in this book provide an understanding of this need and that the strategies recommended provide readers with ways to communicate effectively in their respective fields to their many audiences.

Notes

1. National Commission on the BP Deepwater Horizon Oil Spill and Offshore Drilling, *Deepwater: The Gulf Oil Disaster and the Future of Offshore Drilling: Report to the President* (Washington, DC: Government Printing Office, 2011), 122.

2. Columbia Accident Investigation Board, *Report of the Columbia Accident Investigation Board*, vol. 1 (Washington, DC: Government Printing Office, 2003), 9.

3. Stephen Ambrose, *Nothing Like It in the World: The Men Who Built the Transcontinental Railroad 1863–1869* (New York: Simon and Schuster, 2000), 26.

4. Ibid., 297.

5. Michael Hiltzik, *Colossus: Hoover Dam and the Making of the American Century* (New York: Free Press, 2010), 386.

6. Ibid., 387–89.

7. Columbia Accident Investigation Board, "Columbia Accident Investigation Board Report," vol. 1 (Washington, DC: Government Printing Office, 2003), 106.

8. National Commission on the BP Deepwater Horizon Oil Spill and Offshore Drilling, "Deepwater: The Gulf Oil Disaster and the Future of Offshore Drilling: Report to the President" (Washington, DC: Government Printing Office, 2011), 126.

9. Linda Flower, "Writer-Based Prose: A Cognitive Basis for Problems in Writing," *College English* 41, no. 1 (1979): 19–37.

10. President's Commission on the Accident at Three Mile Island, "Report of the President's Commission on the Accident at Three Mile Island" (1979), 10–11, http://www.threemileisland.org/downloads/188.pdf.

11. Columbia Accident Investigation Board, "Columbia Accident Investigation Board Report," 106.

1 Writing and Reading in the Context of the Environmental Sciences

A CASE STUDY OF THE CHICAGO FLOOD

Introduction

The Chicago flood of 1992 was a man-made environmental disaster,[1] caused by city officials failing to plug a leak in a wall, rather than a natural disaster, such as the lower Mississippi River flood of 2011,[2] caused by the natural forces of melting snow to the north in winter and excessive spring rain. As with so many environmental disasters, the Chicago flood might have been averted had a memo written earlier been heeded.[3] The information Chicago engineers were expected to provide their managers in a request to repair the tunnel leak under the Kinzie Street Bridge in 1992—description of problem, cause, corrective action and estimated cost—is the same today as it was then. The difference is that today the request would have been sent electronically as an e-mail rather than through interoffice mail as a memorandum.

The Chicago Flood

On Monday, April 13, 1992, downtown Chicago came to a standstill. The Chicago River was flooding freight tunnels that had been dug

DOI: 10.5876/9781607324676.c001

underneath the city in the early part of the twentieth century. Water was seeping into the basements of office buildings in the city's financial district, with its stock exchange, commodities market, and Federal Reserve, and into the upscale retail area, where department stores such as Marshall Field's (now Macy's), Burberry, and Neiman Marcus are located. All buildings in the area were evacuated and traffic halted. Downtown Chicago became a ghost town for three days. The city took weeks to pump all the water out. The cost in closed business and ruined inventory ran to approximately $1.25 billion.

The cause of the flood was a leak in the wall of one of the tunnels abutting the river. Several weeks earlier Louis Koncza, the chief engineer for the Bureau of Bridges in the city's Department of Transportation (DOT), had sent a memo to John LaPlante, acting DOT Commissioner, notifying him of the leak and requesting permission to repair the walls. But the wall wasn't repaired before the leak became a flood. Miscommunication between Koncza and LaPlante was one of the reasons.

Beneath the city of Chicago lies a labyrinth of tunnels that were created at the beginning of the twentieth century. The tunnels allowed small train cars to carry coal from the barges that came down Lake Michigan to the basements of the city's buildings above. With the change from coal to oil and gas in the latter part of the twentieth century, the city began to lease these tunnels to cable companies for stringing their cables. In January, four months before the flood, a cable company employee went into the tunnel near the Kinzie Street Bridge to study the situation prior to the company's installing the wire. The employee found water and soil leaking into the tunnel and notified his company, which, in turn, notified the city. A city engineer was sent to investigate but couldn't find a parking place. It took over a month before another employee was sent to check out the report. This time he found a parking place. However, by now the small leak had become a large leak, and the employee found so much damage to the wall that he felt it was unsafe to enter the tunnel. He took photographs and returned. After several meetings to determine what should be done and to estimate the cost for the repair, Louis Koncza was charged with writing a memorandum to the manager of his division, John LaPlante, requesting permission to repair the tunnel (Figure 1.1).[4]

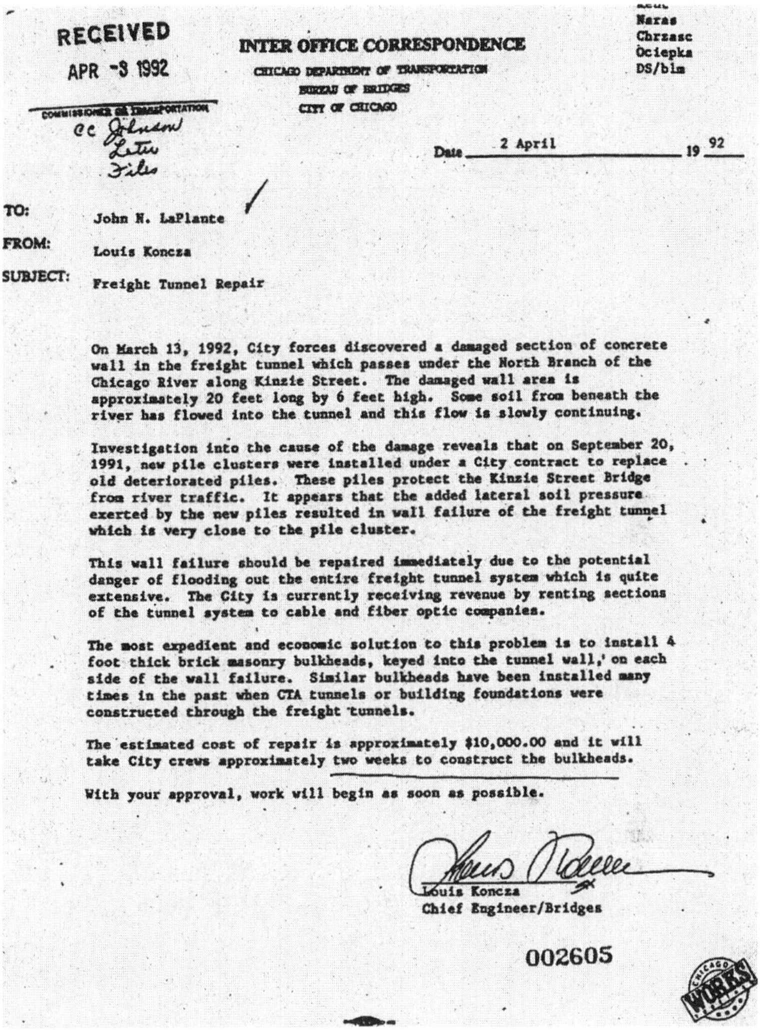

RECEIVED
APR -3 1992

INTER OFFICE CORRESPONDENCE
CHICAGO DEPARTMENT OF TRANSPORTATION
BUREAU OF BRIDGES
CITY OF CHICAGO

Naras
Chrzasc
Ociepka
DS/blm

Date _____2 April_____ 19 92

TO: John N. LaPlante

FROM: Louis Koncza

SUBJECT: Freight Tunnel Repair

On March 13, 1992, City forces discovered a damaged section of concrete wall in the freight tunnel which passes under the North Branch of the Chicago River along Kinzie Street. The damaged wall area is approximately 20 feet long by 6 feet high. Some soil from beneath the river has flowed into the tunnel and this flow is slowly continuing.

Investigation into the cause of the damage reveals that on September 20, 1991, new pile clusters were installed under a City contract to replace old deteriorated piles. These piles protect the Kinzie Street Bridge from river traffic. It appears that the added lateral soil pressure exerted by the new piles resulted in wall failure of the freight tunnel which is very close to the pile cluster.

This wall failure should be repaired immediately due to the potential danger of flooding out the entire freight tunnel system which is quite extensive. The City is currently receiving revenue by renting sections of the tunnel system to cable and fiber optic companies.

The most expedient and economic solution to this problem is to install 4 foot thick brick masonry bulkheads, keyed into the tunnel wall, on each side of the wall failure. Similar bulkheads have been installed many times in the past when CTA tunnels or building foundations were constructed through the freight tunnels.

The estimated cost of repair is approximately $10,000.00 and it will take City crews approximately two weeks to construct the bulkheads.

With your approval, work will begin as soon as possible.

Louis Koncza
Chief Engineer/Bridges

002605

FIGURE 1.1. *Koncza's memo to LaPlante*

The memo was typed on a form for interoffice correspondence that included in the upper right-hand corner the names of the people other than Koncza to whom documents in the DOT were routinely distributed. These people were often involved with some aspect of a project. In this case, one of the men, Ociepka, had been the project manager for the installation of new pile clusters around the Kinzie

Street Bridge, where the leak had begun. Apparently, during the installation, one of the piles had hit the tunnel wall, puncturing it. Another reader, Chrasc, worked under Koncza as his coordinating engineer and would be responsible for coordinating the repair project. Koncza penciled in at the upper left-hand corner the names of three additional people who needed to be informed of the repair work for the leak. Many of these people were eventually fired by then-mayor Richard Daley, Sr.

Koncza, who was busy and had several other problems demanding attention, did not spend much time planning the memo. Instead he wrote it using the same organizational structure that he had used in writing many previous memoranda. He was very much aware of the economic aspect of the tunnel, which brought rental fees to the city, and alluded to this, implying that if the tunnel was not fixed, the city might lose its fees. He did not take much time to read it over and make revisions, other than checking that the facts were correct and then proofreading it for grammatical or spelling errors. On April 2 Koncza sent the memo through interoffice mail to his supervisor rather than delivering it in person as was the convention for matters requiring immediate action.

LaPlante received the memorandum along with a large batch of other interoffice mail the following day, April 3. He constantly received memos describing construction problems and requesting approval to have them repaired. Because this one had been transmitted by interoffice mail rather than in person, it did not appear different from the others.[5]

When LaPlante received the memo in the pile of mail delivered to his desk that Friday afternoon, he simply gathered it along with the entire pile and took it home. On Sunday afternoon, he read through the stack of memos, including the one on the Kinzie Street Bridge.

LaPlante had only recently been appointed to the position in an acting capacity. He was not an engineer but a financial manager. He had little knowledge about the problem discussed in the memo; he wasn't even sure where the Kinzie Street Bridge was. Having no secretary or typewriter available (this was BC [Before Computers]), he handwrote his response on one of his memorandum forms, approving the repair.

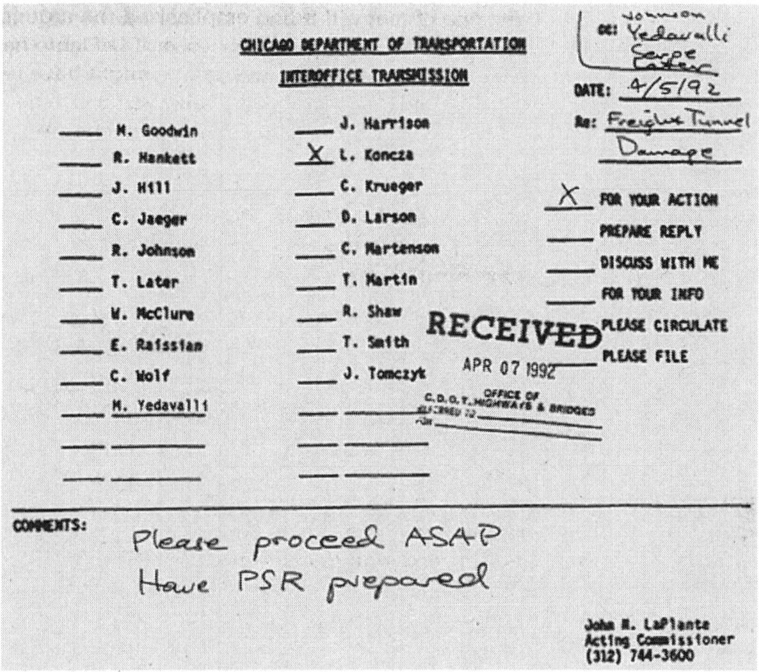

CHICAGO DEPARTMENT OF TRANSPORTATION

INTEROFFICE TRANSMISSION

cc: Johnson, Yedavalli, Serpe, Castor

DATE: 4/5/92

Re: Freight Tunnel Damage

___ M. Goodwin
___ R. Hankett
___ J. Hill
___ C. Jaeger
___ R. Johnson
___ T. Later
___ W. McClure
___ E. Raissian
___ C. Wolf
___ M. Yedavalli

___ J. Harrison
X L. Koncza
___ C. Krueger
___ D. Larson
___ C. Martenson
___ T. Martin
___ R. Shaw
___ T. Smith
___ J. Tomczyk

X FOR YOUR ACTION
___ PREPARE REPLY
___ DISCUSS WITH ME
___ FOR YOUR INFO
___ PLEASE CIRCULATE
___ PLEASE FILE

RECEIVED
APR 07 1992
C.D.O.T. OFFICE OF HIGHWAYS & BRIDGES

COMMENTS:
Please proceed ASAP
Have PSR prepared

John R. LaPlante
Acting Commissioner
(312) 744-3600

FIGURE 1.2. *Approval for a PSR*

Recognizing that the mayor was up for reelection (the political context), he recommended that the repair job be put up for bid. (A PSR [project specification report] is a form for putting out a bid.) He also penciled in the names of several other readers who needed to be kept informed of the project (Figure 1.2).

Returning to work on Monday, LaPlante placed the memo in interoffice mail. It was delivered on Thursday, as the stamp in the upper left-hand corner indicates. It took three more days for Koncza to respond to LaPlante's orders and send a memo requesting that the PSR be processed.

The PSR was prepared and put out for bid. Two bids came in by April 10. Both were over $70,000, approximately, $60,000 over the expected cost (Figures 1.3 and 1.4).

Because the bids were far higher than expected, they were not accepted. It was decided to solicit other bids with the hope that they

PASCHEN CONTRACTORS, *ltd.*

GENERAL CONTRACTORS

April 8, 1992

CHICAGO, ILLINOIS 60647

City of Chicago Department of Transportation Bureau of Bridge Operations, Repair, and Maintenance Mr. Robert Serpe, Director 535 West Grand Avenue Chicago, Illinois 60610

Dear Mr. Serpe:

In response to your request, we offer this proposal and method for bulkheading the damaged service tunnel under the Chicago River.

Due to the potential hazards involved with installing bulkheads at the damaged area locally, we propose to install the bulkheads at the two access shafts.

The work to include:
   Removing muck and debris in the pour area
   Installation of wooden bulkheads (to remain);
   Provide two matts of rebar (dowel into floor);
   Provide 4" pipe and values (each shaft);
   4000 psi concrete;
   Fill tunnel with river water after cure;
   Encase valves with concrete after the tunnel is filled, approx. 2'0" encasement. See the enclosed sketches for detail.

The remaining void area of the shafts would be filled with stone by others. This work to be completed in a period of three (3) weeks for the sum of Seventy-five thousand five hundred dollars ($75,500.00).

It is also understood that adequate area for machinery, concrete trucks , etc. is reasonably available (by land) at each access shaft location. This proposal is based upon the information which you have provided to us, and it does not include a reserve for unanticipated and unforeseeable conditions.

This proposal is subject to changes imposed by your engineering staff once it has had an opportunity to analyze and approve the validity and soundness of the proposed method and plan from an engineering prospective, given all the conditions at and around the site.

If you have any questions pertaining to this proposal, please contact either Mr. Dan Simonides or myself at the above phone number.

Sincerely yours,

Peter Carbonaro Vice President

PC:mr ENCLOSURES

Chicago.

A response to a request for a proposal.

FIGURE 1.3. *Bid to fix the bulkheads by Paschen*

would come in closer to the engineers' estimate. But before other bids could come in, the tunnel wall collapsed and the river flooded the city.

Reading Koncza's Memo

The general consensus of those investigating the flood was that if LaPlante had responded to Koncza's memo by calling for immediate

E. A. C

Cox
ONSTRUCTION COMPANY

Road *Builders* & General Contractors
2424 South Loflin Avenue • Chicago, Illinois 60606
Phone: (312) 738-1333
Fax (312)738-4191

April 10, 1992
City of Chicago Department of Transportation 535 W. Grand Ave. Chicago,
Il. 60610

Attn: Mr. Robert Serpe, Director

Re> Tunnel Repair Work near the Kinzie St. Bridge over the Chicago River

Dear Mr. Serpe.

Regarding the above captioned project, we are pleased to quote on installing a four foot thick brick bulkhead on each *side* of the Kinzie St. Bridge pile cluster location. We assume access will be granted from each side, the Merchandise Mart basement and the Chicago Department of Transportation yard. We also assume access from the Chicago Department of Transportation side will be similar to access on the Merchandise Mart side.

Our lump sum proposal to provide electrical service, labor tools and materials to complete two brick bulkheads is.. $72,500.00

Thank you for the opportunity to quote, and if we can be of any assistance feel free to contact us.

John
Vi
L. COX
Vice-President A. COX COMPANY, INC.

Sincerely

JZC/jat

Figure 3.22
of Chicago.
Letter responding to a request for a proposal.

FIGURE 1.4. *Bid to fix the bulkheads by Cox*

repairs without putting the job out for bids, the flood could proba-bly have been averted. But LaPlante was not able to understand from Koncza's memorandum that he needed to make that decision.

The Chicago flood memo is a prime example of a writer-based rather than a reader-based text and demonstrates what happens when a writer fails to consider readers' processes and styles of reading as well as the context in which the reader will read the message.

Knowledge of the Topic Affects the Reader's Comprehension of the Message

Because LaPlante was not an engineer, he did not think about the extent to which a small hole can become a large hole when it is subjected to water pressure. He also did not consider the damage to electrical wiring and to structural integrity that water can cause when it floods building basements. Nor was he aware of the tunnels or even of the actual site of the Kinzie Street Bridge, and so had no way of recognizing that the financial district as well as the area in which the high-end retail stores were located would be affected if the tunnels flooded. In his memorandum, Koncza needed to provide these explanations to fill in the gap in LaPlante's knowledge of engineering and the area he had only recently been asked to oversee.

Readers' prior knowledge and experience affect their understanding of a topic.[6] To understand a writer's message, readers relate information in a text to their previous knowledge and experience. They then categorize related pieces of information into chunks, sequence information in logical order, and process it both verbally and visually. Comprehending a message is like putting together the pieces of a jigsaw puzzle to create a whole picture. Readers must put all the pieces of information in a document together to *create* a picture of the writer's message.

In order to *create* meaning from the information in a text, readers engage in a three-step process: predicting, reading, and aligning.[7] Readers begin by *predicting* what they will read, based on the situation in which they're reading, the cues they obtain from a document, their knowledge of documents, the topic under discussion, and so on. If they receive a letter, they will predict that they will read information from a client or customer. By glancing at the subject line, they will predict the topic. As they begin to *read* the letter, if the first sentence relates to the topic they predicted, their reading will be *aligned* with their predictions, and they will be able to read the text fluently, without stopping. However, if the first few sentences do not relate to the topic that the readers predicted, then they will stop reading because the text is not aligned with their prediction. They may reread the sentences to try to find a relationship, or they may reconsider their prediction. Either way, they will not have fluency. The infamous memo related to the *Challenger* accident is an example of this major problem (Figure 1.5).

The reader did not read in the first few sentences the answer to the dilemma with which he was faced and for which he was requesting an

MORTON THIOKOL INC
Wasatch Division
9 August 1985 E150/8GR-86-17

Hr. James tf. Thomas, Jr., SA42 George C. Marshall Space Flight Center National Aeronautics and Space Administration Marshall Space Flight Center, AL 35812.

Dear Mr. Thomas :

Subject: Actions Pertaining to SRH Field Joint Secondary Seal

Per your request, this letter contains the answers to the two questions you asked at the July Problem Review Board tclecon.

1. Question: If the field joint secondary seal lifts off the metal mating surfaces during motor pressurization, how soon will it return to a position where contact .is re-established?

Answer: Bench test data indicate that the o-ring resiliency (its capability to follow the metal) is a function of temperature and rate of case expansion. HTI Measured the force of the o-ring against Instron plattens, which simulated the nominal squeeze on the o-ring and approximated the case expansion distance and rate.

At 100^0F. the o-ring maintained contact. At 75^0F. the o-ring lost contact for 2.4 seconds. At 50^0F. the o-ring did not re-establish contact in ten minutes at which time the test was terminated.

The conclusion is that secondary sealing capability in the SRM field joint cannot be guaranteed.

2. Question: If the primary o-ring does not seal, will the secondary seal seat 1n sufficient time to prevent joint leakage?

Answer: MTI has no reason to suspect that the primary seal would ever fail after pressure equilibrium is reached, i.e.,after the Ignition transient. If the primary o-r1ng were to fail from 0 to 170 milliseconds, there 1s a very high probability that the secondary o-r1ng would hold pressure since the case has not expanded appreciably at this point. . If the primary seal were to fail from 170 to 330 milliseconds, the probability of the secondary seal holding 1s reduced. From 330 to 600 milliseconds the chance of the secondary seal holding Is small. This is a direct result of the o-r1ng's slow response compared to the metal case segments as the joint rotates.

FIGURE 1.5. *Response by MTI to questions from NASA Marshall Space Flight Center*

answer: Should the *Challenger* be launched as scheduled. As a result, the reader thought the memo did not provide the information he needed to know: that if the temperature dropped below 50 degrees, the secondary seal would not reseat. (The temperature was well below 50 degrees when the *Challenger* blasted off from Cape Canaveral before it exploded.)[8] A text needs to provide readers with accurate cues for predicting what they will read.

Depending on their knowledge of the topic and field *under discussion*, readers fall into three categories: *experts, generalists*, and *novices*.

Although readers may be experts in their own fields, they may not be experts on a topic discussed in a document. While LaPlante was an expert in financial matters, he knew little about engineering, which was the topic of the memo.

When readers do not have knowledge or experience in a topic, they are far more likely to misinterpret a message. Generalists who have only some knowledge of a field and novices who have no knowledge of the field under discussion need background information that an expert already knows. They also need to have technical terminology defined or replaced by nontechnical words. For instance, when the members of the President's Commission on the Three Mile Island Nuclear Accident wrote their report for the president, Congress, and the general public, they knew they would be writing for readers who were novices in the nuclear field. They spent five pages at the beginning of the account of the accident explaining how a nuclear reactor works so readers could understand what was happening when the problems were described.[9] The commission members also defined such technical terms as "trip" ("a sudden shutdown of a piece of machinery"),[10] terms that are common knowledge to experts in the field but that probably mean nothing or something else to the president, the members of Congress, and the general public, all of whom are novices in the field of nuclear physics.

Recognizing the Purpose of a Message Affects the Reader's Response to a Message

Based on his previous experience in reading similar memos, LaPlante knew that the purpose of the memo was to approve a request so that it could be put out to bid. He responded almost mechanically, without taking the time to consider the implications of the information. For LaPlante to understand that the purpose of the memo was to obtain immediate action without waiting for a bid, Koncza would have had to indicate this in the subject line and first paragraph as well as follow Chicago City Hall transmission protocol by handing the memorandum to LaPlante in person rather than sending it through interoffice mail.

The Context in Which a Reader Reads a Document
Affects the Reader's Interpretation of the Message

Although LaPlante was very much aware of the *economic* advantages of bidding out a project as well as of the *political* necessity of doing so, his lack of knowledge concerning the location of the Kinzie Street Bridge and the areas affected by the tunnels' flooding caused him to misinterpret the importance of putting out a bid on this project. Richard Daley—the present mayor, LaPlante's boss—was up for reelection. He and other city officials had been criticized in the past for giving jobs to their friends, for paying higher fees than necessary for a job, and for failing to provide jobs to minority- and women-owned businesses. By putting the job out for bid, LaPlante was making sure that the cost of the project would be as low as possible and that the mayor's administration would not be criticized. However, the results of the flooding in the financial district and the "million-dollar mile" retail district not only cost the city far more than would have been incurred by either of the high bids, but also cost the mayor political mileage with a disenchanted citizenry.

The city's businesses and citizens became irate at the city's inability to prevent the flood, causing a great deal of political damage in terms of the mayor's candidacy for reelection. Had LaPlante understood these economic and political consequences, he might very well have made a different decision.

Although Koncza commented on the revenue that the city was receiving from renting sections of the tunnel system, he also needed to include information related to the international financial hub along with the multimillion-dollar retail area under which the tunnels ran so that LaPlante could understand the full *economic* consequences of the problem.

Readers' Reading Styles and Patterns
Determine the Information Readers Obtain

Because LaPlante had read many memos requesting approval to fix something, he simply skimmed Koncza's memo, imagining it was similar to others that requested repairs to pavements or potholes. Because it isn't until the third paragraph that Koncza indicates any sense of urgency for repairing the leak, LaPlante may not have even noticed it. In fact, he may not have read that far since he would have had to read through the history of finding the leak, information he didn't really need to know in order to make a decision to approve the repair.

To ensure that LaPlante would read the message rather than cursorily dismiss it, Koncza needed to indicate in the subject line that the problem was critical and repairs needed to begin immediately. Then in the first paragraph he needed to present the problem and his request for immediate action so that, even if Koncza did not read further, he would know what needed to be done and why.

Koncza needed to reorganize his memo so that the information was presented from most to least important rather than chronologically. The request to repair the tunnel wall should have been foremost, and the explanation of why this was necessary should have followed immediately. The history of how the leak was found would have been more appropriately placed toward the end. The estimated cost could have been placed either in the first paragraph with the request or at the end.

The reading styles and behaviors for reading business documents differ markedly from those used to reading a textbook or a piece of literature. Readers do not read page by page or word for word. Rather they skip around in a text; read only the first or last paragraphs of a section; search for specific information; and use a table of contents or index to guide their search. They may read only the abstract or executive summary of a hundred-page report or they may read one or two sections, probably the introduction and conclusion of it. When they receive a memo, they usually look at who it is from and the subject. If they decide they should open the memo, they will quickly read the first paragraph. They may or may not read further.

Readers engage in a variety of reading styles—skimming, scanning, searching, understanding, and evaluating—depending on their purposes and the importance of a document to them.[11] Readers usually simply *scan* a brief, routine memo, or letter to pick out the specific information they need (e.g., the date and time of a meeting, a writer's specific request, information they requested). They look for headings and subheadings, type that jumps out at them because it is in boldface or italics or all-capital letters or a different style, size, or color. They may *skim* through a brief report, reading the first paragraph and the final one, or scan a table of contents to learn the major areas covered in a document. They may *search* through a report to locate information specifically related to their project or division. Readers will only spend the time reading to *understand* and *evaluate* a document if they need the information to work on a project of their own.

Electronic media have affected readers' reading styles. Readers appear to read electronic media more casually than hard copy.[12] They skim the information, seldom stopping or returning to it or even printing it out to read it for understanding. Research has indicated that readers who read on electronic media miss information more often than readers who read a document in hard copy.[13]

Among the many reasons for readers' tendency to skim messages transmitted on electronic media is the proliferation of documents sent via e-mail, the time constraints readers have for reading in the workplace, and readers' inability to spend sufficiently long periods of time concentrating on a single message.

Readers often receive numerous pieces of correspondence during a single day as John LaPlante did. Like LaPlante, readers seldom have time to read the mail as soon as it is delivered or appears on their screens. Usually the mail piles up in an "in" basket on their desk or in their computers. Often they mainly look at their mail, regardless of whether it is hard copy or electronic, first thing in the morning and then sporadically throughout the day, perhaps during the five minutes they are free between appointments or just before going home. Their reading may be interrupted by telephone calls or people stopping at their office or cubicle to talk. In LaPlante's case, he read the memos at home on a weekend when he would rather have been doing other things.

Writing a More Effective Memorandum

Based on the previous discussion of reader-based writing, Koncza might have written his memo more effectively had he written it as follows in Figure 1.6.

Begins with problem. Makes request immediately. Sets up reason to forego bid.	City forces have discovered a damaged section of concrete in the freight tunnel which passes under the North Branch of the Chicago River along Kinzie Street. I am requesting your approval to begin repairs as soon as possible to prevent potential flooding of the financial district and the Million Dollar Mile retail section which the tunnel feeds into. Because of the poor condition of the damaged section and the expectation that the condition will worsen fairly quickly with the pressure of the water, we recommend that the work be done immediately and that the city forego putting the project out for bid.
Effects—Stresses economic consequences.	If the tunnels flood, the potential for structural damage and loss of electricity to the buildings in the affected areas could result in closing off those sections of the city to all traffic and evacuation of all personnel who work in those buildings until the water can be eliminated and the buildings deemed safe. This could cause the City to lose millions of dollars.
Background	In addition, the city is currently receiving revenue by renting sections of the tunnel system to cable and fiber optic companies. Because of the present condition of the tunnel, one company has halted work on stringing cable, causing the City to lose that potential income.
Explanation for repair	The damage was discovered on March 13, 1992. Investigation into the cause of the damage reveals that on September 20, 1991, new pile clusters were installed under a City contract to replace old deteriorated piles. These piles protect the Kinzie Street Bridge from river traffic. It appears that the added lateral soil pressure exerted by the new piles resulted in wall failure of the freight tunnel which is very close to the pile cluster.
Cost	The most expedient and economic solution to this problem is to install 4 foot thick brick masonry bulkheads, keyed into the tunnel wall on each side of the wall failure. Similar bulkheads have been installed many times in the past when CTA tunnels or building foundations were constructed through the freight tunnels. The estimated cost of repair is approximately $10,000.00 and it will take City crews approximately two weeks to construct the bulkheads. With your approval, work will begin as soon as possible.

FIGURE 1.6. *Revision of Koncza's memorandum to LaPlante*

Summary: What Readers Do

1. Readers fall into three categories: novices, generalists, and experts, depending on their familiarity with a topic.

2. Readers use their prior knowledge and experience to help them understand a message.

3. Readers' perceptions of a text are affected by the economic, social, cultural, political, and psychological environment in which they read a document.

4. Readers' purpose for reading a document may differ from the writers' purpose for writing a document.

5. Readers follow a three-step process: predicting, reading, and aligning.

6. Readers engage in a variety of reading styles, including skimming, scanning, searching, understanding, and evaluating.

7. Readers expect documents to follow certain conventions, such as the format of a letter or memorandum, so they can find the information they need when skimming, scanning, or searching a document.

8. Readers' perception of a message is affected by the medium of transmission and its timing.

Notes

1. Richard Daley, "Statement of Mayor Richard Daley, Preliminary Investigation on Flooding P/C," April 14, 1992; Richard Daley, "Statement of Mayor Richard M. Daley, Preliminary Inquiry Update P/C," April 22, 1992; Letter from McDermott, Will and Emery to Janet M. Koran, April 20, 1992, outline of events (FOIA [Freedom of Information Act] 003444); meeting details, Feb. 26, 1992 (FOIA 004906); David Jackson, "In City Hall Memos, Everything Is 'Serious,'" *Chicago Tribune*, April 28, 1992, http://articles .chicagotribune.com/1992-04-26/news/9202070124_1_memo-city-hall -flood; Randall R. Inouye and Joseph D. Jacobazzi, "The Great Chicago Flood of 1992," *Civil Engineering-ASCE* 62, no. 11 (November 1992): 52–55; Patrick Townson, *The Great Chicago Flood of 1992* (1992), http://totse.mattfast1.com /en/politics/political_spew/chiflood.html; Michael Sneed, "The River's Edge," *Chicago Sun Times*, April 16, 1992; David Silverman and William Gaines, "Flood Just a Matter of Inches," *Chicago Tribune*, April 26, 1992, 1.

2. Richard Pallardy, "Mississippi Flood of 2011," *Encyclopedia Britannica* (October 2013), http://www.britannica.com/event/Mississippi-River-flood -of-2011.

3. Matthew L. Wald, "G.M. Illustrates Managers' Disconnect," *Chicago Tribune*, June 9, 2014, B3, http://www.nytimes.com/2014/06/09/business /gm-report-illustrates-managers-disconnect.html.

4. Ellen O'Brien and Lyle Benedict, "1992, April 13: Freight Tunnel Flood," Chicago Public Library, 2005; Karla Bullett, "US Army Corps of Engineers Responds to Strange Flood," Chicago District, archived from the original on June 7, 2008 (May 2002), 37; Leo Dolkart, "The Old Chicago Tunnel," *Midwest Engineer* 7 (December 1963): 22; Bruce G. Moffatt, "The Chicago Freight Tunnels," http://www.mascontext.com/issues/9-network -spring-11/the-chicago-freight-tunnels/ (Spring 2011).

5. Jackson, "In City Hall Memos."

6. Peter Afflerbach, "The Influence of Prior Knowledge on Expert Readers' Main Idea Construction Strategies," *Reading Research Quarterly* 25, no. 1 (1990): 31–46.

7. Judith Langer, "The Reading Process," in *Secondary School Reading: What Research Reveals about Classroom Practice*, ed. A. Berger and H. A. Robinson (Urbana, IL: National Council of Teachers of English, 1982), 39–52.

8. Presidential Commission on the Space Shuttle Challenger Disaster, *The Report of the Presidential Commission on the Space Shuttle Challenger Accident* (Washington, DC: Government Printing Office, 1986); Carl Herndl, Barbara A. Fennell, and Carolyn R. Miller, "Understanding Failures in Organizational Discourse: The Accident at Three Mile Island and the Shuttle Challenger Disaster," in *Text and the Professions*, ed. C. Bazerman and J. Paradis (Madison: University of Wisconsin Press, 1991): 279–305; P. Dombrowski, "The Lessons of the Challenger Investigations," *Professional Communication IEEE Transactions* 34, no. 4 (1991): 211–16.

9. President's Commission on the Accident at Three Mile Island, "Report of the President's Commission on the Accident at Three Mile Island" (1979): 81–89, http://www.threemileisland.org/downloads/188.pdf.

10. Ibid., 173–78.

11. Thomas N. Huckin, "A Cognitive Approach to Readability," in *New Essays in Technical and Scientific Communication: Research, Theory, and Practice*, ed. Paul V. Anderson, R. John Brockman, and Carolyn R. Miller (Farmingdale, NY: Baywood, 1983).

12. Nicholas Carr, *The Shallows: What the Internet Is Doing to Our Brains* (New York: W. W. Norton and Company, 2011).

13. Ferris Jabr, "The Reading Brain in the Digital Age: The Science of Paper versus Screens," *Scientific American* (April 11, 2013): http://www.scientificamerican.com/article/reading-paper-screens/.

2 Effective Discourse Strategies

A Case Study of the 2011 Mississippi Flood

Introduction

The new methods of oil production; alternative power sources, such as wind farms; and the havoc caused by climate change have pitted citizens against industries and forced governments to take steps antagonistic to residents and businesses. Citizens are questioning and increasingly fighting the incursion of fracking into their communities, while the oil and gas industry contends that to date there is no proof of harm caused by this production method. However, recent scientific studies linking earth tremors to fracking appear to indicate a relationship between earthquakes and the disposal of the drilling wastewater. While transportation companies are carrying more oil across country and distribution companies are expanding the size of their pipelines through both rural and urban areas, affected citizens are demanding proof that the expanded routes and lines are safe. Coastal cities face potential flooding of their residential, tourist, and business sections, but residents and business owners contend that they need to continue to build in these areas. All sides need to communicate clearly with each other if they are to reach acceptable solutions.

DOI: 10.5876/9781607324676.c002

Communicating clearly requires writers to make appropriate rhetorical decisions that enable readers to understand their point of view. These decisions relate to focus, content, organization (patterns and sequence of information), and style (language, syntax [sentence structure], voice, and tone) of a document. When these decisions are contextualized in terms of the economic, political, social, cultural, and psychological environment in which a document will be read and when they are reader based, taking into consideration readers' purposes, needs, knowledge, and reading processes and styles, readers are more likely to accept a message.

This chapter discusses the various rhetorical decisions that writers make and shows how some writers have been able to succeed in communicating effectively so that their readers understand their messages and respond to them positively.

The 2011 Mississippi Flood

In the spring of 2011, thousands of square miles of agricultural and residential land were submerged by water that had surged over the banks of the Mississippi River.[1] It became evident that the river would eventually flood major metropolitan areas unless the spillways north of these cities were opened, allowing the water to flow into more sparsely settled farmland. The following case study examines the communications from the Army Corps of Engineers to those affected by the opening of one of these spillways, the Morganza Floodway. The study demonstrates how careful consideration of the rhetorical decisions that writers need to make can result in avoiding a reader's negative responses.

The central Midwest had experienced a great deal of snow during the first few months of 2011. As spring arrived and the snows began to melt, the region experienced heavy rainfall. In fact, the states along the river's tributaries—Missouri, Arkansas, and Tennessee—had their fifth-heaviest rainfall on record.[2] By April the melting snow and rainwater were rushing into the Mississippi's tributaries and into the river itself. The Army Corps of Engineers realized that there was a high probability that the Mississippi River would overflow its banks, creating flooding conditions in major cities, including Baton Rouge and New Orleans.[3] To prevent the flooding of these cities, resulting in major *economic* problems for the metropolitan areas, a *political* problem for the present Obama administration (the dilatory response by the Bush administration to the plight of the residents caught by Hurricane Katrina was still very much present in citizens' minds), and *social* and *psychological* problems for the landowners and residents, the Army Corps of Engineers determined it would be necessary to open the Morganza Floodway, located in central Louisiana.[4] This action would allow the rising waters of the Mississippi to flow into the bottomlands along the river north of Baton Rouge, preventing the flooding of cities further south. However, opening the floodway would displace about 25,000 residents and farmers who lived and owned property along these bottomlands. These people needed to be notified that their land was to be flooded and persuaded to follow procedures to prevent toxic chemicals that may have been stored on their property from spilling into the floodwaters, to relocate livestock, and to evacuate the floodplain.

Since 1954, when the floodway was constructed, landowners and residents had known that their land could be flooded; they had signed a contract with the federal government that stipulated that the government had the right under certain circumstances to open the floodway and that, in such cases, they were to follow certain procedures related to evacuating the area. But the floodway had not been opened since 1973, thirty-eight years prior; many of the present landowners had not even been around at that time.

Because of the long stretch of time since the floodways had been opened, it would have been easy for the landowners to believe that the river would not overflow in their lifetime and to forget about the contract. But the US government realized that the landowners needed to continue to be aware of the possibility of flooding. At the beginning of every year, the government sent a pro forma letter to all landowners to remind them of their contract. However, in February 2011, the Army Corps of Engineers was aware that because of the large amount of snowfall during that winter, the potential for flooding had increased and the landowners might very well be forced to abide by their contract. They were aware of the conflicting psychological impact that the

DEPARTMENT OF THE ARMY
NEW ORLEANS DISTRICT, CORPS OF ENGINEERS
P.O. BOX 60267 NEW
ORLEANS, LOUISIANA

Reply to attention of Real Estate Division

TO: ALL LANDOWNERS, RESIDENTS AND LESSEES IN THE BAYOU DBS GLAISES LOOP, OLD RIVER CONTROL STRUCTURE PROJECT, THE WEST ATCHAFALAYA FLOOD WAY AND THE ATCHAFALAYA BASIN FLOOD WAY AND ALL LANDOWNERS/TRAILER OWNERS OR LESSEE-OPERATORS IN THE MORGANZA FLOODWAY

You are reminded that the property in which you reside, operate and/or are doing business is located within a floodway, or a portion of a floodway, which was developed under appropriate Acts of Congress by the United States Army Corps of Engineers. The function of the floodways is to safely pass the flood waters that may be generated in the Mississippi River and its tributaries to the Gulf of Mexico. The United States Government holds a perpetual right to flood the properties and improvements thereon. The Government's recorded easements provide that the United States shall in no case be liable for damages to property or injuries to persons that may arise from or be incident to operation of the aforementioned floodways.

In the event the operation of any of the floodways is required, public notice will be given through your local civil defense officials, as well as all news media, sufficiently in advance to allow time for evacuation of people and livestock and for removal of personal belongings. Upon receipt of such notice, expeditious action must be taken to protect life and property. Once the floodwaters enter the floodways, the height of the water is estimated to reach between fourteen (14) feet and twenty-five (25) feet above ground elevation depending upon the location of the property within the floodways. As a resident or user of one of the above floodways, you must be aware that it is likely that your home, business and/or personal property could be flooded. In some areas of the floodways, the strength of the currents could thrust improvements from their foundations and carry them through the floodways as debris.

As a user of the floodway, it is your responsibility to minimize environmental contamination during operation of the floodway. Fanners and commercial pesticide applicators that maintain supplies of pesticides must secure and protect these products. In case of high water, measures must be taken to prevent uncontained spills of pesticides, all in accordance with Louisiana Administrative Code Title 7, Pesticides. Also, any property owner that stores or permits storage of hazardous materials, hazardous waste, or NORM (Naturally Occurring Radioactive Material) on their property, must take similar precautions. If you operate any water or gas wells within the floodway, they must be sealed and capped to prevent any contamination from floodwaters.

Owners of trailers in the Morganza Floodway are reminded that in the consent instrument, granting you the right to be in the floodway, you agreed to hold the Government harmless from any and all claims for damages due to flooding, to keep your trailers mobile and capable of being moved on short notice, and to add no permanent appendages to your trailers.

If you have any questions or concerns regarding this matter, you may call our Management, Disposal and Control Branch at 1-800-362-3412, extension 2989.

Sincerely,
Edward R. Fleming Colonel, United
States Army District
Commander

FIGURE 2.1. *Letter 1 from Army Corps of Engineers: model of letter considering readers' needs and wants*

information could have on the residents, causing not only fear of the flooding but also resistance to moving. However, as Ripley indicates, "People will respond to meet a need in a crisis if they know what to do. You give people an opportunity to be a part of something that will make a difference, and they will step up."[5] The annual letter became more significant (Figure 2.1).

DEPARTMENT OF THE ARMY
NEW ORLEANS DISTRICT, CORPS OF ENGINEERS
P.O. BOX 60267 NEW ORLEANS, LOUISIANA 70160-0267

REPLY TO
ATTENTION OF
                    May 6, 2011

Real Estate Division
Management, Disposal, and Control Branch

TO: THOSE WITH IMMOVABLE AND/OR PERSONAL PROPERTY, LIVESTOCK, AND
THIRD PARTY INTERESTS WITHIN THE OLD RIVER CONTROL COMPLEX, MORGANZA
FLOODWAY, LOWER ATCHAFALAYA BASIN FLOODWAY (HWY 190 TO MORGAN CITY
BETWEEN THE EAST AND WEST ATCHAFALAYA BASIN PROTECTION LEVEES), THE
WEST ATCHAFALAYA FLOODWAY, AND THE UPPER POINT COUPEE DRAINAGE
PROJECT

The existing and predicted stages on the Mississippi River indicate that it may be necessary to
operate the Morganza Floodway to handle and safely divert these floodwaters through the Morganza
Floodway. If operated, these waters will ultimately proceed to the Gulf of Mexico through the
Atchafalaya Basin Floodway.

In the event that it becomes necessary to actually operate the Morganza Spillway, public notice
will be given through your State and local emergency officials and other governmental authorities, as
well as major news media, in advance to allow time for evacuation of people and livestock and for
removal of personal belongings, including trailers. Any instructions you have already received or
will receive from State and local emergency officials should be followed. Expeditious action must be
taken to protect life and property.

Once the floodwaters progress through the Morganza and Atchafalaya Floodways to the Gulf of
Mexico, the height of the water could reach between five (5) and upwards of twenty-five (25) feet
above ground elevation, causing widespread flooding and inundation. Possible impacts due to the
water flow currents include: damage to structures, removal of structures from foundations, and
sweeping of other items, such as trailers and livestock, downstream.

Measures must be taken to prevent uncontained spills of pesticides, in accordance with
Louisiana Administrative Code, Title 7, Pesticides. If you store or permit storage of hazardous
materials or waste, or Naturally Occurring Radioactive Material (NORM) on your property, please
take the same precautions. If you operate any type of well within the floodways, it must be sealed
and capped to prevent contamination. In addition, secure any propane tanks in the area.

In summary, please keep advised of instructions from State and local officials as your safety is
our number one priority.

FIGURE 2.2. *Letter 2 from Army Corps of Engineers: model of letter considering readers' needs and wants*

By early May, the Mississippi River had risen far above its normal level. There was a strong possibility that the floodway would need to be opened. Recognizing the *economic* consequences of flooding approximately 46,000 square miles of this highly productive agricultural area and the *social* disruption and *psychological* trauma of relocation to the residents, the Corps sent another letter to residents and landowners to prepare them for the opening of the floodway and to remind them of the steps they needed to take (Figure 2.2). While the letter was *politically* expedient in light of the failure of the government to provide sufficient warning to those in the wake of Hurricane Katrina, it was also an attempt to deflect the citizens from what Sandman calls psychological "outrage" at the government for its role in opening the floodway. By indicating that the relocation was the result of a voluntary decision by those involved to enter into the contract with the government and that the government was providing information to help them prevent the loss of their livestock and the contamination of their wells and land, the writers hoped to obtain the citizens' compliance.[6] In fact, the letter exemplifies the kind of communication recommended in the CAUSE (Confidence, Awareness, Understanding, Solution, Enactment) model of risk communication, especially in terms of the "C" for "Confidence," which may be obtained by legitimizing the concerns of the citizens involved.[7]

Shortly after the letter was mailed, local officials held a series of meetings for local residents to provide them with detailed information for evacuating the region. By May 14, when the floodway was opened, all but the most stubborn of residents had departed the region safely. Although they had not wanted to leave, few had complaints.

Writing the Corps's Letters: Making Appropriate Rhetorical Decisions

The writers of the two letters made rhetorical decisions that resulted in their readers understanding their message and responding to it as the writers had hoped. The writers achieved this result by recognizing their readers' purposes in reading the letters, including content pertinent to the readers' needs and wants, recognizing the context in which their readers would read the letters, and organizing the information so that readers could easily follow the discussion.

Focusing on the Readers' as Well as the Writers' Purposes

Because readers are more willing to read a document if it focuses on their purpose, writers need to take into account the readers' purpose, even if it differs from their own. Both of the letters by the Army Corps of Engineers reflect the readers' as well as the Corps's purposes, even though the writer's purposes for the first letter differ from those of the readers.

The Corps's purposes for letter 1 are (1) to eliminate a negative response if it becomes necessary for the Corps to open the floodway, (2) to cover themselves legally and politically if it becomes necessary to open the floodways, and (3) to prepare residents and landowners for the possibility of such an occurrence. The purposes of the readers who are the residents and landowners affected by the opening of the floodways, are (1) to learn why they are receiving the letter, and (2) to take appropriate action if the information is new and necessary.

By focusing the letter on the original contract and its statement of obligations, the writers were able to reflect the purposes of both the Corps and the readers. The readers' purposes are achieved with the first sentence: (1) they learn why they are receiving the letter and (2) they learn that the information is not new (the first sentence of the first paragraph specifically indicates that this letter is a "reminder"). Even if the readers do not read beyond the first sentence, the Corps has achieved its second purpose—covering itself legally and ethically as well as forestalling complaints that notice was not given in a timely manner. Readers have been reminded of the contract and notified of the possibility of the floodgates opening.

In addition to reminding readers of the contract they signed with the federal government when they purchased their land, the letter fulfills the Corps' other two purposes by reminding them of the ever-present possibility of opening the floodways and their "responsibility" if the floodways are opened.

In letter 2, the readers' and writer's purposes coincide. By now, through the news media, the readers have become aware that the river is flooding. The readers' purposes—(1) to learn whether the flooding will cause the Corps to open the floodway, and (2) to understand the procedures to follow when the floodway opens—are directly related to the writer's purposes—(1) to prepare readers *psychologically* for the opening of the floodway, and (2) to persuade readers to follow procedures to fulfill their responsibilities.

The focus of this letter is on the procedures to follow when the flood-ways are opened. The first sentence of the letter immediately focuses readers' attention on the imminent possibility of opening the floodway while the remaining paragraphs focus on procedures to be followed, thus fulfilling the needs of both writer and readers.

PROVIDING CONTENT READERS WANT AND NEED

Readers *want* information that will help them fulfill their purposes. They are not interested in additional related information or in information that they already know. However, writers may recognize that readers *need* additional information to successfully make a decision or carry out a task. Readers are usually willing to accept this information as long as the information they *want* is included.

Because the focus of the Army Corps of Engineers' first letter is not only to remind readers of their contract and obligations but to cover themselves legally (information readers *need* to know but may not *want* to know), the writer includes the specific terms of the contract along with a listing of the readers' responsibilities, which include a broad outline of the procedures that would need to be followed. In letter 2, the writer refers to letter 1 in case readers have not read it; this reference serves to remind readers of their contract and obligations (information readers *need* but may not *want* to know). Because the focus is on the procedures, the information from letter 1 relating to the procedures is not only repeated but expanded to include information on how readers will learn about the evacuation and the impact of the opening of the floodway on their property (information readers *want* to know).

Determining what readers need to know. In order to base their decisions on the information readers *need* to know, writers consider readers' prior knowledge and experience in relation to a topic. However, it is often difficult to determine whether or not a reader has knowledge about a topic.[8] The bottom line in making a decision regarding the inclusion of information is to "err on the side of caution" and figure that the reader does not have the knowledge; readers can always skim over information they already know.

If readers' prior knowledge does not include the technical vocabulary related to a field or an in-depth understanding of a topic, writers may *need* to include background information, detailed descriptions, and expanded definitions. This information may be provided in a variety of ways. While it may be included in the main text, it may also be placed in a footnote at the bottom of a page or as a note at the back of a chapter or document. It may also be moved to an appendix or a link may be provided to allow the reader to access an Internet site related to the topic. In any of these ways, readers who are unfamiliar with a topic can obtain information about it and those who are knowledgeable can skip it.

In letter 1, the writer provides background information, explaining the purpose for opening the floodway, because readers *need* to know this information in order to understand the reason they are being uprooted. This is information that readers may not remember since the contract has not been activated in almost four decades. However, the writer doesn't include information on the floodway's construction, even though readers may not know it, because that information doesn't help them understand and accept the situation.

In letter 2 the information related to the procedures for removing toxic chemicals is presented only in general terms. The writer does not provide detailed instructions on how to carry out the procedures, assuming the reader already has this knowledge. Because the writer assumes that the readers have read letter 1, he does not outline the terms of the contract, because the readers do not *need* to know those procedures to understand the message.

Determining what multiple readers need to know. Determining readers' prior knowledge, experience, and biases becomes further complicated when multiple readers are involved. Readers' knowledge and experience may differ widely. Both of the letters from the Army Corps of Engineers are read by multiple readers.

Letter 1 (Figure 2.1) is addressed to those who are involved with the land that will be affected. The letter addresses those who own land, those who are lessees of the land, and those who are simply residents in the area as well as trailer owners. Furthermore, the letter covers a relatively large geographic area, involving operators in five different regions. Some of these readers may be absentee landowners who will have a lot invested economically in the land but will not be as psychologically or socially affected as those who are landowners and reside in the area; others may only lease land and so have less invested economically but who will be psychologically and socially affected.

Letter 2 is addressed to many of the same readers but it addresses them in a different role, this time as those who are involved with property on the land, including those with immovable and/or personal property, livestock, or other economic interests and those who are managers or absentee landlords.

When multiple readers are involved, their purposes and needs often differ. In the Corps's first letter, the readers differed psychologically. While some may have been willing to read the entire letter, others were probably unwilling to read information that they had received in previous years, stopping their reading after the first sentence when they realized the information was not new and did not require action. In reading the second letter, however, most of the readers were probably willing, though reluctantly, to follow the orders to evacuate while a small number may not have been.

Readers in the workplace also differ in their roles, positions, and fields. They may be categorized as *primary, secondary, intermediary,* and *peripheral.*[9]

- The *primary* reader is the person or persons for whom a document is being written.
- *Secondary* readers are people affected by a document—marketing managers, comptrollers, and service representatives—who need to design a marketing plan, develop a budget, or repair a piece of equipment.
- *Intermediary* readers are usually supervisors who are responsible for approving a document before it is sent to the *primary* reader.
- *Peripheral* readers, such as a lawyer or news reporter, are persons who are not designated as readers but who may wind up reading a document after the *primary* reader receives it. Because of the use of such omnipresent electronic media as twitter and YouTube, the general public may also read it.

Because documents are never addressed to peripheral readers, writers often forget about them. Yet peripheral readers, such as journalists or attorneys, can cause havoc if they publish a message they have misinterpreted or use a message for legal purposes. In the nuclear power industry, regulations require that all incidents, no matter how insignificant, be reported to the Nuclear Regulatory Commission (NRC) in an Incident Report. These reports are in the public domain and, therefore, available to all citizens, including journalists and attorneys. Several years ago at one of the nuclear utilities, a valve on a sink in a men's bathroom broke and had to be replaced. One of the engineers in reporting the problem in an Incident Report wrote, "a catastrophic [technical term meaning a complete breakdown] valve failure" had occurred. Because the form on which he was required to submit his report had a space requesting "amount of radioactivity released," he had to indicate "zero" even though there was obviously no reason that any radioactivity would have resulted from the broken valve. Several weeks later a local reporter, looking over recent Incident Reports, noted the word "catastrophic." Realizing that the community surrounding the plant had never been notified of this problem, and even though the report indicated that no radioactivity had been released, he wrote up the story, which was printed on the front page of his newspaper. The story was then picked up by the wire services and wound up making national news as well as creating panic in the local community, which questioned the integrity of the company for holding back the information. All of this kerfuffle occurred because the writer of a document had forgotten about peripheral readers and, in describing a broken valve on the sink of the men's room, used the technical term "catastrophic," which a novice peripheral reader misinterpreted.[10]

The letters from the Corps of Engineers (Figure 2.1) were sent to *primary* readers, the residents and landowners in the area, but, prior to sending the letters, the writer sent his drafts to an *intermediary* reader, a supervisor who oversees the Kansas region for the Corps, for approval. Some of the *primary* readers may have passed the letters on to secondary readers, their employees who work the farms. These workers needed to be aware of the procedures outlined in the letters, as they would be responsible for disposing of the toxic chemicals. The letters were also read by *peripheral* readers, the news media, who quoted information from the letters in their local newspapers in order to warn residents and landowners of their impending evacuation as well as to transmit the information to national media.

Organizing Information to Facilitate Readers' Comprehension

Readers want to be able to understand a message quickly and easily. They want information presented so that they do not have to reread a passage or go back to another paragraph or page to figure out how the writer has arrived at a solution to a problem or at a decision among several alternatives. If they aren't willing to take the time to understand a passage that is confusing to them, then they may simply stop reading or they may make assumptions about the information that are incorrect.

A variety of strategies may be used to help readers follow the discussion in a message. These strategies include providing readers with an overall frame for the document, reflecting the purpose of the message in the organizational pattern, sequencing information in a logical order, emphasizing important information, and chunking related information.

Providing readers with a framework for understanding a message. Reading is like putting together a jigsaw puzzle. The reader needs to figure out what message all of the pieces of information are communicating just as the person working a jigsaw puzzle needs to figure out how all of the pieces go together to make the picture on the cover of the box. Once the puzzler finds the four corners and the straight-edged pieces to create the frame, it is much easier to predict where the other pieces go.

Prediction is the first step in a reader's reading process.[11] To understand a message, readers engage in a three-step process: prediction, reading, and alignment.

1. Prediction: Based on their prior knowledge, readers predict what they will read. After reading the first sentence in the Corps' first letter, readers will predict that they are going to read about the location of their land in the government-designated floodway.

2. Reading: Readers read the next portion of a text.

3. Alignment: Readers attempt to align the passage that they have just read with their prediction. If they read what they predicted, as they do in the Corps' first letter, then they have fluency and can continue on to the next passage. If they do not read what they predicted, then they have dissonance and need to reconsider. They may reread the passage in an effort to reinterpret it so that it correlates with their prediction, or they may reconsider their prediction in an effort to create a new one that more closely correlates with the passage, or they may continue reading to see if additional information will help them align their predictions with the text they are reading. If none of these methods provide alignment, then they may stop reading or continue on, misinterpreting the text or failing to comprehend its meaning.

If readers are provided with a frame at the beginning of a document, then it becomes easier for them to understand the information being presented because they can situate the details within the frame. It also provides readers with fluency in reading the text because they can accurately predict what they will read.

The frame for a document should be provided at the beginning of a message and should contain two parts: the statement of purpose/focus and a summary of the information included (the thesis or umbrella statement). If a document is long, it should also contain a third part: a forecast of the major areas covered.

Both letters provide opening umbrella statements that summarize the information to be discussed.

Helping readers locate important information. Readers tend to remember information that is presented first. Furthermore, when readers skim, scan, or search for information, they mainly look at the beginning of a document, at the first paragraph in a section, and at the first sentence in a paragraph. When important information is placed toward the end of a section or paragraph, readers may miss it. For these reasons, the placement of important information in a document is crucial. Placing a frame at the beginning of a document enables the reader to accurately predict the remaining information and to understand how that information is related to the purpose of the document. Important information should also be placed at the beginning of a section of a document just as it should be placed at the beginning of a paragraph. In a sentence, the information to be emphasized should be placed in the independent clause. For example, in letter 1, paragraph 2, the beginning of the first sentence—if the floodways must be opened—is a dependent clause. The important information—that the reader will receive additional information from local civil defense officials—which is the focus of the letter and relates to procedures to follow, is in the independent clause of the sentence.

The example in Figure 2.3 is an introduction to a final report by the Jo Daviess [County, Illinois] Conservation Foundation on its Rivers to Ridges program. The final paragraph provides a forecast of the information contained in the 103-page report.

If a document's organizational pattern reflects a document's purpose, readers can more easily follow the line of argument. Writers can organize texts in a variety of patterns, including general/specific, cause/effect, comparison/contrast, problem/solution, analytic/logical, and most/least important, as well as the conventional scientific research outline, but the writer needs to be careful that the pattern reflects the document's purposes.

If a writer's purpose is to *solve* a problem, then a problem/solution pattern may be appropriate, but if a writer's purpose is to recommend that action be taken to *prevent* a problem, then a cause/effect pattern may be more relevant. In the first situation, readers need to understand the problem in order to understand how the proposed project will solve it. In the second situation, in order to understand the need for the proposed action, readers need to recognize the effects the problem will cause if it is not solved.

Often a chronological pattern appears the easiest way to present information. However, it may not necessarily be the best way to achieve the document's purpose, as the Chicago flood memorandum demonstrates (chapter 1).

The Corps's first letter (Figure 2.1) uses a logical pattern in which the reason for future actions precedes the information on future actions. Letter 2 (Figure 2.2) follows an analytical pattern that covers three top-

INTRODUCTION

Tourism is an essential component of the Northwest Illinois economy. Attracted by a combination of historic character, the Mississippi River, and exceptional driftless topography, visitors spend over $100 million each year just in Jo Daviess and Carroll Counties. The bluffs, prairies, woodlands, and wetlands that help to attract over 800,000 visitors each year have also attracted public and private conservation efforts. This growing collection of sites varies greatly in size, accessibility, and visitor amenities. Though some, such as the Mississippi Palisades, are well known and well used, others remain well kept secrets, or simply aren't yet ready for visitors.

In 2006, the Jo Daviess Conservation Foundation (JDCF) was awarded funding from the Hamill Family Foundation to inventory these diverse sites and prepare a plan to promote their use and enjoyment. JDCF selected planning consultant MSA Professional Services, Inc to assist with plan development, and recruited a project steering committee comprised of representatives of many of the site owners. The planning process features a series of Steering Committee meetings at the JDCF offices in Elizabeth, beginning in September 2007 and concluding in May 2008.

September 6, 2007	Kickoff
November 13, 2007	Visitors' Profile & Market Demand review;
	Site Inventory & Assessment review
January 20, 2008	Branding & Wayfinding Plan review
March 30, 2008	Branding & Wayfinding Plan review
May 15, 2008	Marketing Plan review, complete plan review

The document is broken into six sections: market research, site assessments, branding, wayfinding, marketing, and action plan.

FIGURE 2.3. *Helping readers follow a line of reasoning:* Rivers to Ridges, final report *by the Jo Daviess [County, IL] Conservation Foundation*

ics: (1) information on the status of the flooding, (2) information on how the readers will be notified, and (3) actions readers will need to take.

Helping readers recognize relationships among data. A document needs to be both coherent and cohesive so that readers can follow a writer's line of reasoning from section to section, paragraph to paragraph, and sentence to sentence. By chunking related ideas, by using transitions between ideas, and by repeating key words from one passage to the next, writers can help readers recognize how various ideas are related to each other and to a document's main focus and purpose.

- *Chunking.* Because one of the ways in which readers comprehend texts is by relating the various pieces of information to one another, related information needs to be chunked together. Readers do not want to jump back and forth between pages or screens, trying to put together pieces of information that are scattered in different areas of a message. Thus, all background information should be in a single section or paragraph of a text. In the Corps's first letter (Figure 2.1), the information related to the contract is contained in paragraph 1, all the information related to the opening of the floodway is in paragraph 2, and all the information related to the readers' responsibilities is contained in paragraph 3, while information related specifically to trailer owners is contained in paragraph 4.

- *Transitions.* When documents are coherent and cohesive, readers can follow the writers' line of reasoning and recognize the relationship between pieces of information. Transitions, such as "on the other hand," "next," "because," "in addition to," and "therefore," help readers understand when information contrasts with other information, follows other information in a sequence, is the effect of previous information, is expanding on previous information, and is the result of previous information, respectively.

- *Repetition.* Repeating a word from a previous paragraph or sentence in a new paragraph or sentence helps provide topic coherence.[12] While word repetition is frowned upon in creative writing and often academic documents, it is necessary in technical/scientific writing. When a word is repeated from one paragraph to the next, readers are reassured that they are reading about the same topic as they were in the previous paragraph. Nor is there any confusion as to which topic the writer is referring. The word "floodways" is repeated numerous times in the Corps's first letter.

Readers can easily put the pieces of information together and follow the writer's logic in both of the Corps's letters. Related information is placed in paragraph chunks. Transitions—such as "also," in letter 1 and "as well as," and "in addition" in letter 2—provide cohesion within sentences. Temporal transitions, such as "once the floodwaters enter/ progress," found in letters 1 and 2 respectively, provide sequential cohesion. Repetition of the word "floodways" provides topic coherence in both letters: in letter 1 the word is used in the final sentence of the first paragraph and is then repeated in the opening phrase of the second paragraph. This pattern is repeated in each of the ensuing paragraphs, providing coherence to the whole message.

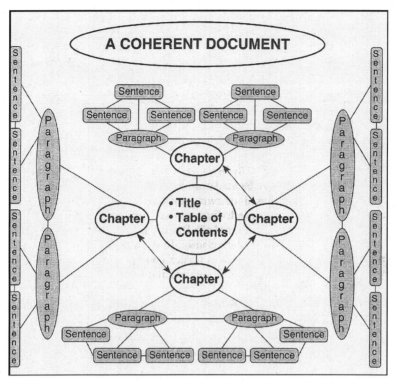

FIGURE 2.4. *Model of a coherent and cohesive document*

A diagram of a coherent document in which all of the parts are related to create a single message is modeled in Figure 2.4.

Helping Readers Read Fluently

Sentence structure can affect readers' fluency and comprehension. When a sentence is dense, containing too many ideas, it becomes difficult for readers to understand the relationships between the various ideas. Sentence 1, below, is long (forty-six words) and dense.

> In 1988, when Genesis Research Corporation approached Duquesne Light to cosponsor development of an SO_2 emission control technology that Genesis had conceived, Duquesne Light contacted CQ Inc. (CQ), then the Electric Power Research Institute's (EPRI's) Coal

Quality Development Center, to perform an independent review of the proposed technology.

This sentence contains the following eight kernels (propositions):

1. Genesis approached Duquesne.
2. The approach related to the cosponsoring of a project.
3. The project involved SO_2 emission control technology.
4. The company that conceived the technology was Genesis.
5. Duquesne contacted EPRI.
6. Duquesne requested an independent review.
7. The review concerned the SO_2 technology.
8. Duquesne requested the review when Genesis approached it.

The preceding sentence is composed of two major concepts, each contained in a clause: (1) Genesis Research Corporation approached Duquesne Light Company (dependent clause) and (2) Duquesne contacted Electric Power Research Institute (independent clause). By breaking the sentence into two separate sentences, the information can be made easier to read and comprehend. Readers will find it is easier to understand the two rather than the original single sentence.

Sentence 1: *In 1988, Genesis Research Corporation approached Duquesne Light to cosponsor development of an SO_2 emission control technology which Genesis had conceived.*

Sentence 2: *Duquesne Light contacted CQ Inc. (CQ), then the Electric Power Research Institute's (EPRI) Coal Quality Development Center, to perform an independent review of the proposed technology.*

Sentences in technical documents should be a maximum of twenty-five words.

Sometimes long sentences are inevitable because the ideas behind them are complex. However, readers will be able to read them more easily if they follow the basic sentence structure of subject, verb, object. When this order is interrupted, it becomes more difficult for readers to follow a line of reasoning. In the first sentence below, it is difficult to connect the subject and predicate, whereas in the second sentence the reader has no problem in determining the connection.

1. The pipeline, which is routed through a portion of Canada where there are many moraines, could be more likely to have structural problems if upheavals, caused by volcanic activity, though none are presently expected, occur, and should be rerouted.

2. The pipeline should be rerouted because there are many moraines in the portion of Canada through which it is routed, and thus it is more likely to have structural problems if there are upheavals caused by volcanic activity, though none are presently expected to occur.

The second sentence is clearer because the subjects of the various clauses ("pipeline" and "upheavals") are not separated from their predicates ("could be more likely," "occur," and "should be rerouted"), as they are in the first sentence.

Summary

A. FOCUS
- The focus of a document should include the readers' as well as the writer's purposes.
- The focus should relate to the readers' and writer's purposes.

B. CONTENT
- Content should reflect the readers' previous knowledge and experience.
- Content should relate to the focus and purpose of the message.
- Content should include information that the readers do not know but needs to know to understand and accept the text.
- Content should *not* provide information that the readers already know.
- Content should reflect the readers' previous knowledge and experience.
- Content should reflect readers' biases.
- Content should reflect the wants and needs of multiple readers.

C. ORGANIZATION
- A text should be organized to facilitate readers' comprehension.
- Information in a text should be organized and sequenced according to the readers' and the writer's purposes.

- The focus of a document should be related to its purposes.
- A frame should be provided at the beginning of a document so readers can predict what they will read.
- A frame should include a statement of the purpose of the document and a sentence summarizing the information contained in the document.
- A forecast that lists the main topics in a document should be included in the frame if the document is long.
- Information should be organized so that readers can accurately predict what they will read.
- Important information should be placed at the beginning of a document, section, and paragraph so readers will not overlook it.
- Related information should be chunked together.
- The main idea of a sentence should be placed in the independent clause.
- If readers are unfamiliar with or unknowledgeable about a topic, the background, explanations, and descriptions should be placed first to familiarize readers with a topic.
- Information should be sequenced in logical order, using one or a combination of the following organizational patterns: alphabetical, chronological, analytical, sequential, comparison/contrast, problem/solution, cause/effect, most/least effective, most/least important.

D. READABILITY

- Readers should be able to understand a text easily and fluently.
- If readers may be unfamiliar with the topic, details should be included and technical terms defined. Definitions should be related to a document's purpose.
- Transitions and repetition should be used to help readers recognize relationships between ideas.
- Terminology and sentence structures should be those with which readers are familiar.
- Readers should be able to understand a text easily.
- Complicated sentence structures should be avoided.
- Sentences should average fifteen to twenty words and should have a maximum of twenty-five words.

Notes

1. Douglas Carlson, Marty Horn, Thomas Van Biersel, et al., "Atcha-falaya Basin Inundation Data Collection and Damage Assessment Project" (Baton Rouge: Louisiana Geological Survey, 2011), http://data.dnr.louisiana.gov/ABP-GIS/ABPstatusreport/Report_of_Investigation_12-01web.pdf; Paul Rioux, "Morganza Floodway Open to Divert Mississippi River Away from Baton Rouge, Louisiana," *Times Picayune*, May 4, 2011, http://www.nola.com/environment/index.ssf/2011/05/morganza_floodway_opens_to_div.html.

2. National Oceanic and Atmospheric Administration, "United States Flood Loss Report—Water Year 2011," 2013, http://www.nws.noaa.gov/hic/summaries/WY2011.pdf.

3. Mark Schleifstein, "Mississippi Flooding in New Orleans Area Could Be Massive If Morganza Floodway Stays Closed," *Times Picayune*, 2011, www.nola.com/weather/index.ssf/2011/05/army_corps_fears_massive_flood.html.

4. US Army Corps of Engineers (2011), "Morganza Floodway Proposed Clarifications to Standing Instructions," http://www.mvn.usace.army.mil/Missions/MississippiRiverFloodControl/MorganzaFloodwayOverview.aspx.

5. Amanda Ripley, *The Unthinkable: Who Survives When Disaster Strikes—and Why* (New York: Three Rivers Press, 2009), 211.

6. Peter M. Sandman, *Responding to Community Outrage: Strategies for Effective Risk Communication* (Fairfax, VA: American Industrial Hygiene Association, 1993).

7. Katherine E. Rowan, "Earning Trust and Productive Partnering with the Media and Public," *Consortium of Social Science Associations* (2004), http://www.cossa.org/seminarseries/risk_and_crisis.htm.

8. Jeremy Roschelle, "Learning in Interactive Environments: Prior Knowledge and New Experience" (1995), https://www.sri.com/sites/default/files/publications/imports/RoschellePriorKnowledge.pdf.

9. Carolyn Boiarsky, *Technical Writing: Contexts, Audiences and Communities* (Boston, MA: Allyn and Bacon, 1993), 120–24. Other authors have categorized readers similarly. David McMurrey, "Audience Analysis: Just Who Are These Guys?" (2015), https://learn.saylor.org/mod/page/view.php?id=5540 l; Mary B. Coney, "Technical Readers and Their Rhetorical Roles," *IEEE Transaction on Professional Communication* 35, no. 2 (June 2, 1992): 58–63; David Pearson, "Readers and Contexts of Use," http://www.pearsonhighered.com/samplechapter/0205632440_ch3.pdf.

10. Story in local Tennessee newspaper, around 1986.

11. Judith Langer, "The Reading Process," in *Secondary School Reading: What Research Reveals about Classroom Practice*, ed. A. Berger and H. A. Robinson (Urbana, IL: National Council of Teachers of English, 1982), 39–52.

12. M.A.K. Halliday and Ruqaiya Hasan, *Cohesion in English* (New York: Pearson Education Ltd., 1976).

3 Effective Persuasive Strategies

CHEAP, AVAILABLE COAL POWER VS.
A CLEAN ENVIRONMENT

Introduction

Environmental issues are seldom black and white, and solutions often require compromise on both sides. Everyone wants clean air, but communities also need industries to employ their citizens. Power to provide everything from lights in a home to X-rays in a hospital to electricity in a factory is as essential as clean water for drinking. Those on either side of these issues communicate constantly with the general public as well as with their professional communities in an effort to persuade them that their opinions on the topic are the appropriate ones.

Readers involved with these issues not only read in an economic and political context, they also read in a cultural, social, and psychological one. Their attitudes toward the various topics related to environmental risks are often emotional, reflecting deeply held beliefs that Kahn suggests are integral to their cultural beliefs.[1] Readers are also influenced by the people with whom they work, as well as by those with whom they socialize.[2] While some people read information in order to make a decision, many read in order to be supported in the opinions they have already adopted.[3] When an argument supports an idea to which they

DOI: 10.5876/9781607324676.c003

are opposed, they may become defensive, respond negatively, or refuse to respond at all.[4] This reaction occurs especially when highly controversial topics with inherent risks are discussed, such as regulations for safer railcars carrying light oil, the reduction of carbon pollutants, and the purchase of wetlands to build malls. Arguments related to such topics, if they are to be effective, need to relate directly to the reasons readers have for holding their opinions.

Continued Use of Coal as a Source of Energy

One of the most controversial topics is that related to coal. While it is one of the most readily available and cheapest means of energy,[5] it is also one of the major sources of carbon dioxide (CO_2).[6] Attempts to reduce the use of coal draw arguments from those states in which coal mining is a major source of employment as well as from those industries that use coal as their source of power and from countries, such as China, that have used coal to move into the twenty-first century. Numerous attempts have been made to find a way to continue to use coal as a source of power while simultaneously eliminating its deleterious effects on the climate and on people's health. One of the methods being attempted is that of capturing and storing the gas.[7] The US Department of Energy began an initiative in the 1990s to introduce this method to industry by providing funding for several power companies around the country. The results have been mixed and to date few companies have adopted it. Simultaneously, the government has attempted to enact regulations to lower the amount of CO_2 by enforcing the use of already-proven procedures, such as the Utility MACT Rule that sets standards to limit acid gases and other toxic pollutants from power plants. The rule would affect approximately 1,400 coal- and oil-fired plants and could cost as much as $9.6 billion by 2016.[8] Industry has fought these regulations, arguing that they are costly to industry and will drive up prices for consumers.[9] The rule is presently in limbo. The Supreme Court voted against part of the rule in June 2015. The Environmental Protection Agency has since clarified one part of the rule, but as of January 30, 2016, it has not been enacted.

Recognizing Readers' Biases

Before beginning to draft a message, writers need to determine the claims and supporting evidence that are most closely related to their readers' concerns.

The excerpts of documents in Figures 3.1 and 3.2 are each concerned with the continued use of coal, but are written by organizations with opposing viewpoints. The documents are posted on their organizations' respective websites. They are read by a broad range of readers, from experts to novices; from engineers to environmentalists, business leaders, and interested citizens; and from those who agree with the documents' respective viewpoints to those who don't. The first excerpt (Figure 3.1) is posted by America's Power, an organization sponsored by the American Coalition for Clean Coal Electricity (ACCCE), a partnership of industries involved in producing electricity from coal. The purpose of the excerpt is twofold: (1) to persuade its members that it is succeeding in pressuring Congress to retract regulations that the members do not want, and (2) to persuade other readers that the legislation is harmful. The second example, an excerpt of a progress report from the US Department of Energy (Figure 3.2), is part of a section describing the Clean Coal Power Initiative undertaken by the DOE.[10] Its purpose is to persuade readers that its initiative is a good one.

Reflecting Readers' Attitudes

In the excerpt in Figure 3.1, the writer wants to persuade readers to support the Senate's action to eliminate the EPA Utility MACT regulation. The writer focuses the argument on the economic needs of the readers as individuals and as members of a community. Recognizing that readers will not want to spend more for their electricity, the argument focuses on the potential increase in the cost of power if the regulations remain. Paragraph 2 introduces a second argument that focuses on job loss, a problem that was especially relevant in 2012, as the United States slowly emerged from the Great Recession. Again this argument is directly related to readers' needs for employment for themselves and for their community.

AMERICA'SPOWER

Follow us on:

SEARCH   SB   ADVANCED SEARCH «

AFFORDABLE   ABUNDANT   CLEAN COAL TECHNOLOGY   GET PLUGGED IN   NEWSROOM   ISSUES & POLICY   TAKE ACTION

ACCCE Statement on U.S. Senate Effort to Restrain EPA's Job-Destroying Regulation

Thursday, February 16, 2012 Category: Press Room Utility MACT is the Wrong Policy at the Worst Time

WASHINGTON - The United States Senate today took the first step in setting aside a new EPA regulation on America's coal-fueled electricity industry that would unnecessarily drive up energy costs for millions of American families and businesses. Senator James Inhofe (R-OK), Ranking Member of the Senate Committee on Environment and Public Works, filed a joint resolution of disapproval under the Congressional Review Act regarding the Environmental Protection Agency's Utility MACT Rule. EPA also calls the rule the MATS rule.

"Congressional action is essential to stop this heavy-handed new regulation by EPA that will needlessly drive up energy prices for all Americans and destroy jobs," said Steve Miller, president and CEO of the American Coalition for Clean Coal Electricity. "With half of Americans now devoting more than 20 percent of their family budget to energy costs, EPA is making energy much more expensive with the most costly regulation ever imposed by the agency on the coal-fueled electricity industry. Given the fragility of America's economic recovery, that's the wrong policy, at the worst possible time."

The effects of rising energy costs are the subject of a recently released report, "Energy Cost Impacts on American Families." Some of the major findings are that -

☐ Energy costs have almost doubled for the average family and are eating up a disproportionate share of low- and fixed-income families' budgets.
☐ Energy cost burdens are greatest on the poorest families.
☐ Minority families are particularly burdened by higher energy costs.
☐ Lower- and fixed-income senior households are among those most vulnerable to energy price increases.
☐ Electricity is a relative bargain among energy products. This is due, in part, to the utility industry's reliance on affordable coal.
☐ EPA regulations drive up electricity prices. Electricity price increases over the past two decades are due in part to costs associated with meeting clean air and other environmental standards.

The full study is available at:
http://www.americaspower.org/sites/default/files/Energy_Cost_Impacts_201...

EPA's Utility MACT regulation, which was finalized in late December 2011, would lead to the closure of many coal-fueled power plants. A comprehensive analysis by National Economic Research Associates found that the proposed Utility MACT Rule and other finalized and pending EPA regulations for power plants using coal could destroy an average of 183,000 jobs every year from 2012- 2020 and increase electricity and other energy prices by $170 billion.

"America's coal-based electricity industry is grateful to Senator Inhofe and other Members of Congress who understand coal's importance to the American economy now and well into the future," said Miller. "We urge strong, bipartisan support for this resolution in order to protect consumers and jobs."

FIGURE 3.1. *Argument against EPA ruling: "ACCCE Statement on US Senate Effort to Restrain EPA's Job-Destroying Regulation" (http://americas power.org/accce-statement-us-senate-effort-restrain-epas-job-destroying -regulation)*

In the excerpt in Figure 3.2, the writer recognizes the readers' interest in the world's ecosystems and health and focuses on reduction of acid rain in the beginning of the excerpt and pollution in the second half of the statement.

ENERGY.GOV

CLEAN COAL POWER INITIATIVE

"Clean coal technology" describes a new generation of energy processes that sharply reduce air emissions and other pollutants from coal-burning power plants.

In the late 1980s and early 1990s, the U.S. Department of Energy conducted a joint program with industry and State agencies to demonstrate the best of these new technologies at scales large enough for companies to make commercial decisions. More than 20 of the technologies tested in the original program achieved commercial success.

The early program, however, was focused on the environmental challenges of the time - primarily concerns over the impact of acid rain on forests and watersheds. In the 21st century, additional environmental concerns have emerged - the potential health impacts of trace emissions of mercury, the effects of microscopic particles on people with respiratory problems, and the potential global climate-altering impact of greenhouse gases.

With coal likely to remain one of the nation's lowest-cost electric power sources for the foreseeable future, the United States has pledged a new commitment to even more advanced clean coal technologies.

Building on the successes of the original program, the new clean coal initiative encompasses a broad spectrum of research and large-scale projects that target today's most pressing environmental challenges.

The Clean Coal Power Initiative (CCPI) is providing co-funding for new coal technologies that can help utilities cut sulfur, nitrogen and mercury pollutants from power plants. Also, some of the early projects demonstrated ways to reduce greenhouse emissions by boosting the efficiency by which coal plants convert coal to electricity or other energy forms.

In January of 2003, **eight projects were selected under the first round CCPI solicitation**, of which two were withdrawn. Of the remaining six projects supported by the first round of the CCPI, one was discontinued before award, two were discontinued during project development, and three were successfully completed.

In October of 2004, **four projects were selected from the second round CCPI solicitation**. Two projects have withdrawn, one project is complete and the remaining project is under construction. The ongoing project with Southern Company Services is expected to enter commercial operation by mid-2015.

In 2009 and 2010, **six projects were selected from third round CCPI solicitations**. These projects utilize carbon capture and storage technologies and/or beneficial reuse of carbon dioxide. Due to an additional $800 million of funding added to the CCPI Program through the American Recovery and Reinvestment Act of 2009 (Recovery Act), Round Three was conducted through two separate solicitations. Of the six overall projects selected, three projects from Round Three (HECA, Summit and NRG) are still active..

In January of 2003, **eight projects were selected under the first round CCPI solicitation**, of which two were withdrawn. Of the remaining six projects supported by the first round of the CCPI, one was discontinued before award, two were discontinued during project development, and three were successfully completed....

In October of 2004, **four projects were selected from the second round CCPI solicitation**. Two projects have withdrawn, one project is complete and the remaining project is under construction. The ongoing project with Southern Company Services is expected to enter commercial operation by mid-2015.

In 2009 and 2010, **six projects were selected from third round CCPI solicitations**. These projects utilize carbon capture and storage technologies and/or beneficial reuse of carbon dioxide. Due to an additional $800 million of funding added to the CCPI Program through the American Recovery and Reinvestment Act of 2009 (Recovery Act), Round Three was conducted through two separate solicitations. Of the six overall projects selected, three projects from Round Three (HECA, Summit and NRG) are still active.

FIGURE 3.2. *Argument for clean coal technology: update on clean coal initiative, 2012 (http://energy.gov/fe/science-innovation/clean-coal-research /major-demonstrations/clean-coal-power-initiative)*

Reflecting Readers' Concept of Valid Evidence

Reflects readers' biases. Readers are not only more apt to be persuaded by claims that relate to their own attitudes, but also to accept the validity of evidence that comes from sources that reflect their own biases.[11]

For example, one of the objectives of the Alaskan government is to preserve the populations of moose and caribou. Sportspeople who write a letter to persuade state government officials to continue to permit the aerial shooting of wolves will want to focus their letter on the claim that relates aerial shooting to the preservation of moose and caribou populations. On the other hand, representatives of conservation organizations—such as the Defenders of Wildlife, who want to halt the practice of aerial shooting—will want to focus on the claim that the moose and caribou populations are declining because of the practice of aerial shooting, not because of wolves, who do not go after the moose and caribou. The writer may also want to cite research that has been conducted to support his or her views.

Regardless of which side the writer is on, the research cited as evidence will need to have been conducted in Alaska if the readers are to accept it. If the research is conducted in other states, readers may counterargue that the conditions in Alaska differ and that the evidence, therefore, isn't valid.

Reflects conventions of readers' professional communities. Readers who are experts in their field often judge the validity of a document by a writer's ability to refer to the appropriate authorities, follow the appropriate procedures, and use the technical terminology, conventional patterns, styles, and format appropriate to their professional communities. Some readers may accept support for an argument from Wikipedia, but other readers may only accept evidence based on research in a scientific journal. Because readers of both groups would accept a scientific journal, the writer would do well to use it rather than Wikipedia.

Both of the excerpts in Figures 3.1 and 3.2 cite research studies to support their respective claims. The ACCCE excerpt cites two studies to support its claims. Those who agree with the writer's biases will probably perceive the results of the study as support for the claims presented. In addition, those who place their pocketbooks above their environmental concerns will agree with the claims. However, one of the studies, "Energy Cost Impacts on American Families," was conducted by the same organization as the one for whom the writer is writing, and the other was conducted by National Economic Research Associates, a consulting firm owned by a large corporate conglomerate. Those who are undecided or against the bill may perceive the reported results as invalid because the one study uses a researcher who is associated with the project rather than an outsider as required by the criteria for valid research accepted by the science community. They may also view the results of the other study as invalid because it was conducted by an organization with biases that support coal production.

The US Department of Energy (DOE) excerpt has the same problems with validity. Those who agree with its premise will probably not question its conclusions, but those who disagree with the idea or who are members of a scientific field may question the conclusions. Letter 2 refers to results of projects conducted by its own staff and is, therefore, biased toward clean coal production. In fact, it does not indicate the criteria on which the projects were judged successful nor does it provide the name of a specific report in which these results are reported, all of which are conventions accepted by the scientific community.

Every professional field has its own set of conventions that govern the way in which information is to be presented if it is to be considered valid. These conventions are described in style guides that are developed and published by a community. The conventions cover everything from citing quoted information, to writing an abstract, to setting up a title page. The community of literary critics follows the conventions of the Modern Language Association (MLA), which requires that a writer cite an author's name and the page number of the material in referencing an excerpt or idea in a major published paper. The applied engineering community follows the conventions of such professional organizations as the American Society of Mechanical Engineers (ASME) and the Institute of Electrical and Electronic Engineers (IEEE), which require that writers cite an author and the year an article or book is published rather than a page number in referencing material in a document. The applied science community, such as the National Association of Environmental Professionals (NAEP) and the American Academy of Environmental Engineers and Scientists (AAEES), follows a similar style. In writing documents in the environmental sciences, a writer may need to follow the style guide of the specific field in which the report is written. For example, a biologist may need to use the style established by the the American Society of Plant Biologists (ASPB) if readers are to be persuaded to accept a study's results. Many organizations print their own style guides, which can often be obtained from their publication departments.

Helping Readers Follow a Line of Reasoning

When readers are provided with a summary statement at the beginning of a message and when that message follows a logical organizational pattern that is coherent and cohesive, they are more likely to accept a concept. Cause/effect and problem/solution organizational patterns are often the most appropriate for a message in which the sender wants to persuade the responder to do something or think in a specific way.

The first paragraphs in the excerpts from both the ACCCE and the DOE contain umbrella statements that summarize the information that will be presented in their documents, providing readers with a frame for the messages.

Because the purpose of the DOE document (Figure 3.2) is to provide readers with information on improvements in the department's programs, the writer uses a cause/effect organizational pattern, explaining in a chronological sequence the reason for (*cause of*) a new DOE program and then discussing the *effect* of the changes, using an analytical pattern in which the various projects of the initiative are discussed. The ACCCE document (Figure 3.1) follows a general (the policy) to specific (the deleterious effects of the policy) pattern to reflect its purpose in informing readers of the problems of the policy.

Review

- Arguments should consider readers' prior knowledge, experience, biases, and fields so that readers will consider the claims valid.

- Conventions for citing sources and references should be appropriate to the readers' community. Claims with supporting evidence should relate to the readers' purposes.

Notes

1. According to Dan Kahan, "Culture is prior to facts . . . what citizens believe about the empirical consequences [of a risk] . . . *derives* from their cultural worldviews." Dan Kahan and Donald Braman, "Cultural Cognition and Public Policy," *Yale Law & Policy Review* 24, no. 47 (2006): 148.

2. Kahan suggests that a process that accounts for the polarization of beliefs related to topics such as climate change or coal is "cultural cognition": "the influence of group values—ones relating to equality and authority, individualism and community—on a person's "risk perceptions and related' beliefs." He suggests that "a major cause of political conflict over the credibility of scientific data" is "protective cognition," in which people "tend to dismiss evidence of environmental risks, because . . . such evidence would restrict activities they admire." Kahan, "Fixing the Communications Failure," *Nature* 463, no. 7279 (2010): 296–97.

3. Kahan suggests that "cultural cognition" "causes people to interpret new evidence in a biased way that reinforces their predispositions." Ibid., 296–97.

4. According to Kahan, people reject arguments because of "cognitive-dissonance avoidance," the inclination to avoid beliefs that may "force one to renounce commitments and affiliations essential to one's identity." Kahan, "Cultural Cognition and Public Policy."

5. Robert Bryce, "Dirty but Essential—That's Coal," *Los Angeles Times*, July 27, 2012, http://articles.latimes.com/2012/jul/27/opinion/la-oe-adv -bryce-coal-epa-climate-20120727; Deborah Tedford, "Why We Still Mine Coal" (Washington, DC: National Public Radio, August 8, 2010), http:// www.npr.org/templates/story/story.php?storyId=125694190. On coal's availability, see Institute for Energy Research, "Coal," *Encyclopedia* (Washington, DC: Institute for Energy Research, 2013), http://instituteforenergyresearch. org/topics/encyclopedia/coal/. On coal's low cost, see Ben Geman and Nathaniel Gronewold, "Coal Fired Power Plants Will Need Better Carbon Capture and Storage Technology," *Scientific American* (February 12, 2009), http://www.scientificamerican.com/article/coal-fired-power-plants-carbon-capture/; Richard Conniff, "The Myth of Clean Coal," *Yale Environment 360*, June 3, 2008, http://e360.yale.edu/feature/the_myth_of_clean_coal/2014/.

6. *Sourcewatch*, "Environmental Impacts of Coal," March 2015, http://www.sourcewatch.org/index.php/Environmental_impacts_of_coal; Union of Concerned Scientists, "Coal Is a Dirty Energy Source," Cambridge, MA, January 2014, http://www.ucsusa.org/clean_energy/smart-energy-solutions/decrease-coal/.

7. American Public Power Association, "Electric Generating Utility (EGU) Mercury MACT Rule" (February 2012), http://www.publicpower.org/files/PDFs/EGUMACTRuleFeb2012IB.pdf; Bipartisan Policy Center, "Assessment of EPA's Utility MACT Proposal," March 2011, http://bipartisanpolicy.org/wp-content/uploads/sites/default/files/Q&A%20Assessment%20of%20MACT%20Rule.pdf.

8. Hal Rogers, "Rogers Statement on EPA Issuance of Utility MACT Rule," December 21, 2011, http://halrogers.house.gov/news/documentsingle.aspx?DocumentID=273432; Margaret Ryan, "EPA MACT Rule Released: Coal Plants Set for Closure as Blackout Risks Cited," *Breaking Energy*, December 21, 2011, http://breakingenergy.com/2011/12/21/epa-utility-mact-rule-released-coal-plants-set-for-closure-as-b/; Carolyn M. Brown and Robin B. Thomerson, "United States Supreme Court Reverses Utility MACT Rule," *National Law Review*, June 29, 2015, http://www3.epa.gov/mats/actions.html.

9. Sourcewatch, "Environmental Impacts of Coal," March 2015, http://www.sourcewatch.org/index.php/Environmental_impacts_of_coal.

10. National Mining Association, "Clean Coal Technology" (Washington, DC: Department of Energy, 2013), http://www.nma.org/pdf/fact_sheets/cct.pdf; Department of Energy, "The Clean Coal Technology Program," February 12, 2013, http://www.fe.doe.gov/education/energylessons/coal/coal_cct2.html.

11. Kahan, "Fixing the Communications Failure"; Daniel M. Kahan and Donald Braman, "Cultural Cognition and Public Policy" (2006), http://papers.ssrn.com/sol3/papers.cfm?abstract_id=746508; Brendan Nyhan, Jason Reifler, Sean Richey, et al., "Effective Messages in Vaccine Promotion: A Randomized Trial," *Pediatrics* (Elk Grove Village, IL: Journal of the American Academy of Pediatrics, 2014), http://pediatrics.aappublications.org/content/early/2014/02/25/peds.2013-2365; Joe Keohane, "How Facts Backfire," *Boston.com*, July 11, 2010, http://www.boston.com/bostonglobe/ideas/articles/2010/07/11/how_facts_backfire/; David McRaney, "The Backfire Effect," *You Are Not So Smart*, June 10, 2011, http://youarenotsosmart.com/2011/06/10/the-backfire-effect/.

4 Communicating with Electronic Media

CASE STUDIES IN THE *COLUMBIA* SHUTTLE BREAKUP
AND THE BP/*HORIZON* OIL RIG EXPLOSION

Engineers and scientists engaged in such large-scale projects affecting the environment and ecological infrastructure as fracking, light oil pipeline expansion, and offshore oil rig construction are constantly making on-the-job decisions that could have dire consequences. Decisions related to complex questions are often complicated by the exigencies of the actual situation. For example, the BP engineers who were on the Horizon oil rig prior to the explosion needed to know whether they could safely use fewer centralizers than required in constructing an offshore oil rig but they were geographically separated from the managers who could determine the answer to the question. In the case of the BP/*Horizon* Gulf oil rig explosion, some of the managers and crew were on land while others were on the rig. In the case of the *Columbia* shuttle accident, engineers and managers were communicating among various divisions within NASA as well as with the astronauts on the shuttle in space.

Solutions to the kinds of dilemmas posed by *Horizon* and *Columbia* require informed judgments based on deep knowledge of a subject. People who work on solving these kinds of risky situations need to

DOI: 10.5876/9781607324676.c004

be able to discuss alternatives and collaborate to arrive at acceptable decisions. The men and women involved with the BP and *Columbia* tragedies communicated much of the time via electronic media. Some of these communications were effective; others were not. This chapter examines the major communications that occurred in both cases and studies the reasons some succeeded and others failed.

Communicating via Electronic Media

All of the rhetorical strategies recommended for writing effective letters and memoranda also apply to writing electronic messages. Writers need to keep in mind at all times the readers, the readers' purpose, and the context in which the readers read a message. However, in electronic communication, writers need to use additional rhetorical strategies because of the format and the mode of transmission.

E-mail and text messages are a hybrid form of communication in which the letter and memorandum have coalesced, and written communication has converged with the telephone.[1] While e-mail and text messages continue to possess some of the rhetorical aspects of the letter and memorandum as well as the telephone, they are conducted in a technological context.[2] As a result of these convergences, new formats and conventions have developed.

The e-mail default format is based on the conventions of a memorandum ("To," "From," "Subject," and "Date"), but the opening and sign-off follow the conventions of a letter, requiring a salutation at the beginning, a closing (Thanks, Regards, Best), and the writer's name/ signature at the end. The senders and receivers of electronic messages assume dual roles in the same way that those involved in telephone conversations alternate between the roles of listener and speaker. Conversation threads related to the same subject create a chain of e-mails or text messages with participants responding sequentially to each other (Figure 4.1).

While e-mail and text messages possess some of the characteristics of each of the media they have subsumed, they also lack certain aspects. One of the major differences lies in the amount of content included in an e-mail or text message. Electronic messages tend to be information

Kathleen Webster

From:	Maureen Bardusk [mb@maureenbardusk.com]
Sent:	Monday, January 11, 2010 10:44 PM
To:	websters@galenalink.com
Cc:	webster@galenalink.com
Subject:	Fwd: Rivers to Ridges

okay, all systems go.
Maureen

-----Original Message-----
From: Paul Tobin [mailto:ptobin@wc314.org]
Sent: Monday, January 11, 2010 01:52 PM
To: 'Maureen Bardusk'
Subject: RE: Rivers to Ridges

That is correct and look forward to seeing you. If you could arrive around 11:45 that would work out best. We meet in the conference room and order our meal ahead of time at The Subway on site and they are delivered to us around 12:30. My cell is 815 275-6339 if you need anything else.

Paul Tobin

-----Original Message-----
From: Maureen Bardusk [mailto:mb@maureenbardusk.com]
Sent: Monday, January 11, 2010 12:43 PM
To: Paul Tobin; Paul Tobin (External)
Cc: websters@galenalink.com
Subject: Rivers to Ridges

Hello, Paul

This is to confirm that Kathleen and Jim Webster will make a presentation re Rivers to Ridges on Tuesday, January 26, at noon at Land of Oz for the Mt. Carroll Rotary Club. Please confirm that we have the correct date, time, and location.

Thank you for the invitation.

Maureen Bardusk
Rivers to Ridges

mb@maureenbardusk.com
815.777.1242

FIGURE 4.1. *Example of an e-mail chain with conventions derived from letters and memoranda*

poor.[3] While letters and memos can transmit long, detailed, and complex information, e-mails and texts are mainly limited to the length of a single screen. Long e-mails require scrolling, and readers usually prefer to stop at the end of the screen if they are not motivated to continue.[4] Because attachments are usually long and can be cumbersome to open, especially on a smartphone, they are often saved to be read at a later time.[5] Electronic messages also tend to communicate less emotional content than that of telephone conversations; emoticons for enthusiasm, approbation, or sympathy do not exist.[6]

Another major difference is that although e-mails and texts, like telephones, allow participants to engage in the threads of a conversation in real time, in most cases people e-mailing and texting do not. However, they often perceive that they do. In most cases reading and writing occur sequentially, not simultaneously; the actions are asynchronous. In an electronic message, a reader usually does not read the words as a writer writes them, while in a telephone conversation a listener hears the words as the speaker says them and can interrupt at any time to request more information, to ask for clarification of a point, or to disagree with a concept. Usually, even with the ability to send and receive voice messages, a person sends an e-mail or text and then goes on to do something else rather than wait for a reply.

Reading on Electronic Media

Almost all technical and scientific communication today is conducted via electronic media,[7] and the media are expanding. Messages are being read on smartphones, tablets, and notebooks as well as on computers and netbooks, and watches and eyeglasses are also beginning to provide a means of communication.[8] For those between the ages of eighteen and thirty-nine, the smartphone has become the primary device for reading e-mails.[9] While the processes and styles of reading messages on electronic media remain basically the same as those for reading traditional written texts, readers have modified their reading patterns in order to adapt to the new platforms. Manufacturers are also continuously adapting their devices to meet readers' needs for fluency and easy comprehension.

To improve reading on a screen, smartphones are becoming larger and the default font is changing. Research has indicated that reading on a screen is easier with a sans serif rather than serif font, with Ariel 12 point or Verdana at 10 or even 9 point, or Cambria at 11 point, as the most readable.[10]

Research also indicates that people read their e-mail in an "F" pattern. They read the very beginning of a message (often just the heading) and then skim vertically down the left margin, in the process scanning the beginning of the second paragraph, then returning to vertically skim the remainder of the screen before deciding to read further or to stop reading. If important information does not fall along this path, readers are likely to miss it.[11]

Not only have the media on which a message is delivered expanded, but so have the number of messages that readers are receiving. Users check their smartphones for e-mail approximately 150 times a day.[12] The typical corporate e-mail user receives about 105 e-mails per day.[13] Employees are expected not only to respond to a message, but to respond to it immediately, creating what I shall call, for want of a better name, Electronic Communication Stress Syndrome (ECSS).[14] This stress has resulted in messages being read and responded to quickly, without much thought.[15] Although reading and responding to a message requesting attendance at a meeting can be done in this fashion, reading and responding to a message that concerns a critical engineering decision often cannot be.

Reading e-mail and text messages can be likened to a pilot's description of flying a plane: hours of boredom interrupted by moments of sheer terror. Most electronic messages are common everyday communications about meetings and questions and responses related to basic work responsibilities. However, every now and then, a message is sent that is of major importance. The e-mails that were sent during the *Columbia* shuttle accident were among those.

COLUMBIA SHUTTLE ACCIDENT

On February 1, 2003, *Columbia* was heading toward home when the left wing began to break up.[16] Over the next few minutes, the shuttle disintegrated, with pieces of its fuselage scattered from California to Texas. It was determined that the breakup was caused by a piece of insulating foam that separated from the external tank during launch and penetrated the tiles of the thermal protection system, creating a hole in the leading edge of the left wing. During reentry, the extremely high temperatures of the air penetrated the wing through the hole, melting the interior structure of the wing and causing the breakup similar to the way in which the planes hitting the Twin Towers on 9/11 caused the structures to melt.

Columbia was one of five shuttles that were sent into space between 1981 and 2011 with a twofold purpose: (1) to carry scientific experiments for private and governmental institutions, as well as industrial payloads, such as satellites, into space; and (2) to carry astronauts, equipment, and supplies to the space station. A total of 135 missions were flown during that thirty-year period. The funding for the space program had decreased over the years. By 2003 Congress was pushing NASA to become self-sufficient through its payloads as originally planned. Reductions in funding were causing NASA to cut back on maintenance and to hide problems or, at the very least, find shortcuts for fixing them. In this environment, employees were required to assume a posture of success and to refrain from noting any negative aspects or challenges.

From the beginning of the shuttle program, there had been a problem with the Spray On Foam Insulation (SOFI), pieces of foam that were attached by hand, one by one, to the outer frame of the orbiter to protect it from the heat it encountered upon reentry into the atmosphere. At one point, an enterprising mechanic even tried using the chewing gum in his mouth as an adhesive to hold the tile in place. This makeshift adhesive actually caused the tiles to stick better in the short term, but once the shuttle was in space, it did not hold.

Although pieces of foam had been flying off the shuttle over the years, this one appeared larger than previous ones and seemed to have hit the craft with more force than the others. NASA was not aware of the hit until day 2, when the Intercenter Photo Working Group at the Marshall Space Center in Huntsville, Alabama, received a high-resolution film taken during launch that showed "debris impacting" the underside of the left wing. But they could not see the exact location of the hit. Although there are usually a dozen cameras set up around the globe to photograph the shuttle's passage, some of them were not operating for this launch; thus, there were no photographs taken of the impact. CAIB (Columbia Accident Investigation Board) found that "of the dozen ground-based camera sites used to obtain images . . . , five are designed to track the Shuttle from liftoff until it is out of view."

However, because of a limited view and dense atmospheric conditions, two sites were unable to catch the debris event. Of the remaining three sites, "the first site lost track of *Columbia* on ascent, the second site was out of focus—because of an improperly maintained lens—and the third site captured only a view of the upper side of *Columbia*'s left wing. . . . Over the years, it appears that due to budget and camera-team staff cuts, NASA's ability to track ascending Shuttles has atrophied."[17]

Because foam had become detached during many of the previous flights, even hitting the shuttle during several of the launches, managers were not worried. However, the members of the NASA Photo Working Group were concerned. This piece of foam appeared larger than the others and had apparently hit with more force. In an attempt to obtain specifics about the location of the hit and the extent of the damage, the group sent a request to the Department of Defense to photograph the shuttle. But they went through the back door. When the head of the mission, Linda Ham, learned about the request, she voided it.

The photo group was not the only group working on this problem. The NASA Debris Assessment Team was also worried about the damage and felt the need for a better picture of the hit. Their chair, Rodney Rocha, sent an e-mail to two of the managers of his own department at Johnson Space Center in Houston, requesting assistance from upper management to obtain photographs of the area where the debris hit the shuttle (see Figure 4.2).

Rocha never received a reply to his e-mail. If his readers used the "F" pattern for reading e-mails, then it is possible that they missed the purpose of his letter—to get them to request assistance in photographing the damage—which is located in the second paragraph. Because the first sentence simply iterates an agreement by participants at a meeting, and the remainder of that paragraph (skimming the left margin of it) appears to be only a reiteration of the problem of which the readers are already aware, readers may have decided that there was no new information and stopped reading before the second bar of the "F."

In addition, Rocha failed to transfer effective discourse strategies from traditional written texts to his electronic texts. His message is writer rather than reader based. He fails to consider the priorities of

--- Original Message--

From:	ROCHA, ALAN R. (RODNEY) (JSC-ES2) (NASA)
Sent:	Tuesday, January 21, 2003 4:41 PM
To:	SHACK, PAUL E. (JSC-EA42) (NASA); HAMILTON, DAVID A. (DAVE) (JSC-EA) (NASA); MILLER, GLENN J. (JSC-EA) (NASA)
Cc:	SERIALE-GRUSH, JOYCE M. (JSC-EA) (NASA); ROGERS, JOSEPH E. (JOE) (JSC-ES2) (NASA); GALBREATH, GREGORY F. (GREG) (JSC-ES2) (NASA) **Subject:** STS-107

Wing Debris Impact, Request for Outside Photo-Imaging Help

Paul and Dave,
The meeting participants (Boeing, USA, NASAES2 and ESS, KSC) all agreed we will always have big uncertainties in any transport/trajectory analyses and applicability/extrapolation of the old Arc-Jet test data until we get definitive, better, clearer photos of the wing and body underside. Without better images it will be very difficult to even bound the problem and initialize thermal, trajectory, and structural analyses. Their answers may have a wide spread ranging from acceptable to not-acceptable to horrible, and no way to reduce uncertainty. Thus, giving MOD options for entry will be very difficult.

Can we petition (beg) for outside agency assistance? We are asking for Frank Benz with Ralph Roe or Ron Dittemore to ask for such. Some of the old timers here remember we got such help in the early 1980's when we had missing tile concerns.

Despite some nay-sayers, there are some options for the team to talk about: On-orbit thermal conditioning for the major structure (but is in contradiction with tire pressure temp, cold limits), limiting high cross-range de-orbit entries, constraining right or left had turns during the Heading Alignment Circle (only if there is struc. damage to the RCC panels to the extent it affects flight control.

Rodney Rocha
Structural Engineering Division (ES-SED)
☐ ES Div. Chief Engineer (Space Shuttle DCE)
☐ Chair, Space Shuttle Loads & Dynamics Panel

Mail Code ES2

FIGURE 4.2. *Rocha requests that photographs be taken of the area where the foam hit the shuttle. The e-mail is an example of the merged conventions of a letter and memorandum.*

his readers, which are related to the success of the flight, not to such technicalities as improved data. Rocha's emphasis should have been on the effect of the photographs on the final outcome of the flight and the lives of the astronauts, as he implies in the final paragraph, instead of on the effects of the photographs in conducting a data analysis that he discusses in his first paragraph. Thus, the important information—the effect on a reentry decision—is placed at the end rather than the beginning. And, in fact, that information appears to contradict prevailing thought on the part of management.

By formulating this request in the form of a question, "Can we petition (beg) for outside agency assistance?" rather than as a declarative statement, "We are requesting outside agency assistance," or even, "We need outside agency assistance," Rocha further diminishes his authoritative tone, leaving plenty of opportunity for a nonresponse from his readers.

---Original Message--
From: Robert H. Daugherty
Sent: Monday, January 27, 2003 3:35 PM
To: .   CAMPBELL, CARLISLE C, JR (JSC-ES2) (NASA)
Subject: Video you sent

WOW!!!
I bet there are a few pucker strings pulled tight around there!
Thinking about a belly landing versus bailout....................(I would say that if there is a question about main
gear well burn thru that its crazy to even hit the deploy gear button...the reason being that you might
have failed the wheels since they are aluminum..they will fail before the tire heating/pressure makes
them fail..and you will send debris all over the wheel well making it a possibility that the gear would
not even deploy due to ancillary damage...300 feet is the wrong altitude to find out you have one gear
down and the other not down...you're dead in that case)
Think about the pitch-down moment for a belly landing when hitting not the main gear but the trailing
edge of the wing or body flap when landing gear up...even if you come in fast and at slightly less pitch
attitude...the nose slapdown with that pitching moment arm seems to me to be pretty scary...so much
so that I would bail out before I would let a loved one land like that.
My two cents.
See ya,
Bob

From: "CAMPBELL, CARLISLE C., JR (JSC-ES2) (NASA)"
To: "'Bob Daugherty'"
Subject: FW: Video you sent
Date:Mon, 27 Jan 2003 15:59:53 -0600
X-Mailer: BInternet Mail Service (5.5.2653.19)

Thanks. That's why they need to get all the facts in early on-such as look at impact damage from the spy telescope. Even then, we may not
know the real effect of the damage.

The LaRC ditching model tests 20 some years ago showed that the Orbiter was the best ditching shape that they had ever tested, of
many. But, our structures people have said that if we ditch we would blow such big holes in the lower panels that the orbiter might break up.
Anyway, they refuse to even consider water ditching any more~I still have the test results[Bailout seems best.

FIGURE 4.3. *Excerpt from a discussion between two engineers occurring over a two-day period, January 27 and 28, and concerning two topics: a video and tile damage. It is an example of an effective e-mail chain.*

Several days after Rocha sent his message, without knowing the actions the photo group engineers had taken, the engineers on the Debris Assessment Team requested a photograph of the affected area. The message was intercepted by a middle manager who responded for Ham, indicating that she had already refused the request. In the meantime, NASA staff on the ground was working around the clock. E-mails, such as the one in Figure 4.3, were continually being sent between the various divisions and between the engineers and managers in an effort to determine the risk the astronauts faced and, if necessary, to find a means to rescue them. Figure 4.3 is an example of an effective e-mail chain in which two people have a dialogue concerning the consequences of the potential damage caused by the foam. Despite the lack of time for reading and responding to e-mails, this reader responds within twenty-four minutes, far less than the twenty-eight hours that usually occur between messages.[18] The relatively quick response may be related to several factors. The reader receives a positive emotional response—"Wow!"—to his sending of a video, which probably motivated him to continue reading the message. He also receives the information he wanted: a reaction to the video.

As the days grew closer to the time the astronauts would have to return to Earth, the engineers continued to work feverishly on a solution. Robert Dougherty at Langley, Virginia, Air Force Base, who had not been specifically assigned to work on the Debris Assessment Team, sent an e-mail to one of the members of the team, David Lechner, offering a scenario different from the ones they were working on and suggesting they consider his along with the others (Figure 4.4). His scenario is eerily close to what actually happened. While NASA officially contends that the e-mail was read and considered, there are those who believe the e-mail was disregarded, a likely scenario.

If the readers at NASA had followed the "F" pattern,[19] they would have missed the main point by the time Dougherty gets to it. The beginning is personal and contains far more information about his present position and state of mind than is relevant. In the first paragraph Dougherty concentrates on his own failings and feelings as if he were posting on Facebook. He rambles for the first four very long sentences. He does not get to the point of the message until the eighth line down, and it is stuck in the middle of a very long first paragraph where it is basically buried in an independent clause.

Had the engineers been able to secure the photographs they wanted, they would have seen the damage that the impact had caused. Whether or not that recognition would have made a difference in the final results depends on whom you talk to. Some say a rescue shuttle might have been able to save the astronauts if a decision had been made at the very beginning of the crisis in time to get it up; others say it couldn't have been done.

Writing for Electronic Media

Electronic media have made reader-based messages even more necessary. Brevity to avoid scrolling and the placement of the most important information at the beginning of a message to accommodate readers' "F" formation reading pattern have become essential. Additional strategies

Hi David,

I talked to Carlisle a bit ago and he let me know you guys at MOD were getting into the loop on the tile damage issue. I'm writing this email not really in an official capacity but since we've worked together so many times I feel like I can say pretty much anything to you. And before I begin I would offer that I am admittedly erring way on the side of absolute worst-case scenarios and I don't really believe things are as bad as I'm getting ready to make them out But I certainly believe that to not be ready for a gut-wrenching decision after seeing instrumentation in the wheel well not be there after entry is irresponsible. One of my personal theories is that you should seriously consider the possibility of the gear not deploying at all if there is a substantial breach of the wheel well (color mine). The reason might be that as the temps increase, the wheel (aluminum) will lose material properties as it heats up and the tire pressure will increase. At some point the wheel could fail and send debris everywhere. While it is true there are thermal fuses in the wheel, if the rate of heating is high enough, since the tire is such a good insulator, the wheel may degrade in strength enough to let go far below the 1100 psi or so that the tire normally bursts at. It seems to me that with that much carnage in the wheel well, something could get screwed up enough to prevent deployment and then you are in a world of hurt. The following are scenarios that might be possible...and since there are so many of them, these are offered just to make sure that some things don't slip thru the cracks...I suspect many or all of these have been gone over by you guys already:

1 People talk about landing with two flat tires...I did too until this came up. If both tires blew up in the wheel well (not talking thermal fuse and venting but explosive decomp due to tire and/or wheel failure) the overpressure in the wheel well will be in the 40 + psi range. The resulting loads on the gear door (a quarter million IDS) would almost certainly blow the door off the hinges or at least send it out into the slip stream. ..catastrophic Even if you could survive the heating, would the gear now deploy? And/or also, could you even reach the runway with this kind of drag?
2.. The explosive bungles...what might be the possibility of these firing due to excessive heating?
If they fired, would they send the gear door and/or the gear into the slipstream?
3. What might excessive heating do to all kinds of other hardware in the wheel well...the hydraulic fluid, uplocks, etc? Are there vulnerable hardware items that might prevent deployment?
4. If the gear didn't deploy (and you would have to consider this before making the commitment to gear deploy on final) what would happen control-wise if the other gear is down and one is up?
(I think Howard Law and his community will tell you you're finished)
5. Do you belly land? Without any other planning you will have already committed to KSC. And....

Admittedly this is over the top in many ways, but this is a pretty bad time to get surprised and have to make decisions in the last 20 minutes. You can count on us to provide any support you think you need.

Best Regards,

Bob

FIGURE 4.4. *Dougherty's e-mail is writer based, fails to adhere to basic discourse strategies for effective reader-based documents, and does not allow for the reader's "F" reading pattern.*

can ensure communication is effective when it occurs on electronic media. Some of the e-mails and text messages sent during the three months prior to the BP/*Horizon* Gulf oil rig explosion were effective; others weren't.

The BP/*Horizon* Gulf Oil Rig Explosion

At 9:40 PM on April 20, 2010, an explosion ripped through the *Horizon* oil rig, anchored approximately fifty miles off the Louisiana coast in the Gulf of Mexico.[20] The blast resulted in the largest accidental marine oil spill in US history, with over 200 million gallons of oil (4 million barrels) covering approximately 29,000 square miles (an area about the size of South Carolina). Of the 126 workers on board the rig, 11 died in the blast and 17 were injured. The company has paid over $42.2 billion in criminal and civil settlements. Environmental effects were devastating. There were reports of tuna and amberjack exposed to the oil showing deformities of the heart and other organs. In the *Proceedings of the National Academy of Sciences*, scientists reported that chemicals from the spill were found in migratory birds as far away as Minnesota; pelican eggs contained "petroleum compounds and Corexit [used to disperse oil in a spill]"; and PAHes (Polynuclear Aromatic Hydrocarbons) from oil probably caused "disturbing numbers" of mutated fish. Fifty percent of shrimp were found lacking eyes and eye sockets. Fish with oozing sores and lesions were first noted by fishermen in November 2010, and a definitive link was found between the death of a Gulf coral community and the spill.[21]

Preparation of an oil well for drilling occurs in two stages: (1) construction of the well, which requires a large rig on which the workers construct the well and ready it for drilling, and (2) the actual drilling for oil, which requires a smaller rig. The explosion occurred at the conclusion of stage 1. The construction of the well was almost complete; workers were capping off the well in order to replace the *Deepwater Horizon* rig with a smaller rig and begin the drilling. The operation involved three companies: BP Plc, which owned the well; Transocean Ltd., which was responsible for leasing the large rig to BP; and Halliburton Co., which was responsible for the cementing process.

By April 2010, the project was about six weeks behind schedule and more than $58 million overbudget. Management was anxious to complete construction so that the well could begin operation and start making money.

The men working on the rig were as anxious as management to complete the job and move off the rig. They had been following a demanding schedule of twelve-hour workdays, seven days a week, for three weeks at a stretch, eating and sleeping on board the rig. Brian Morel, one of those who had designed the well, exclaimed in an e-mail, "This has been a nightmare well."

From the beginning, the project was plagued with problems. BP leased the large rig *Marianas* from Transocean Ltd. in October 2009 to begin the first stage of the Macondo oil well. However, thirty-four days after the *Marianas* had been brought in, it was damaged by Hurricane Ida and had to be replaced with the *Deepwater Horizon* rig. Work on the new rig did not begin until several weeks into 2010.

Problems continued to plague the project. According to one of the men on the rig, "At times, the drill got stuck. Many times it 'kicked,' meaning gas was shooting back through the mud at an alarming rate . . . I've never been on a well with that high of gas coming out of the mud."[22]

A BP engineering reorganization resulted in "delays and distractions."[23] A change in management occurred; decisions were made, revised, and remade. One official demanded that workers "replace heavy mud, used to keep the well's pressure down, with light seawater to help speed a process that was costing an estimated $750,000 a day."[24] A decision to replace the usual ingredients for a cement slurry with a new mixture that had never been tested was made. In early April, a fracture occurred in the rock formation that was to be drilled, and engineers realized that they could not drill as deeply as they had planned (20,200 feet below sea level) if they were to maintain "well integrity and safety."[25] They were forced to stop almost 2,000 feet short of their goal at 18,360 feet. The final problem involved centralizers (metal strips for keeping the cement even in the well) that are used to cement the annulus area (the space in the well made for the drill) in the final phase of the construction. This process required the use of twenty-one centralizers. However, it was discovered that the supplier had sent only six. When a new supply arrived, it was determined that the new ones were a different, unacceptable kind. Rather than wait until fifteen more of the same type could be located and sent, BP decided to go with just the six.

The report to the president concluded, "Many of the decisions that BP, Halliburton, and Transocean made that increased the risk of the

Macondo blowout clearly saved those companies significant time (and money)."[26]

By mid-April the well was ready to be capped off so the *Deepwater Horizon* could be replaced by a smaller rig for production. This process, called "temporary abandonment," requires two tests—a positive and a negative test—to ensure that the cement is secure. The positive test, which checks that nothing is seeping *out* of the well, was successful. Engineers began the negative test at 5 PM on April 20. The object of this test is to check that nothing is leaking *into* the well. Psi must remain at 0 for two pipes, the drill pipe and the kill line, which are inserted into the well.

The test was not successful. Results indicated a psi of approximately 1,400 in the drill pipe. The engineers made several revisions to the equipment, but the psi on the drill pipe remained at about the 1,400 level. One of the Transocean engineers, Jason Anderson, who had been on the rig since the beginning, suggested that the result was caused by what he called the "bladder effect" and indicated that he had seen this kind of result previously. (Anderson was one of those who died in the explosion.)[27] Don Vidrine, a BP rig supervisor, insisted that a second test be conducted on the kill line. However, while the psi on the kill line dropped to 0, the psi on the drill pipe remained at 1,400. (Vidrine has since been indicted for "seaman's manslaughter." He is accused of "botching a key safety test and disregarding abnormally high pressure readings that were glaring signs of trouble before the blowout."[28]) Despite the test failure for the drill pipe, the well site leaders decided to accept the kill line results and at 8 PM began the process of plugging the well. One hour and forty minutes later, the explosion occurred.

In examining the communication between the time the engineers discovered the lack of centralizers and the explosion, it becomes relatively clear that many of the messages were ineffective. With e-mail and texting at the finger (or thumb) tips of the engineers involved with the BP/*Horizon* Gulf oil rig, communication among BP, Transocean, and Halliburton engineers and managers, and between those on shore and those on the rig, should have been quick and simple. It was not. The National Commission on the BP Deepwater Horizon Oil Spill and Offshore Drilling Report to the President found that the root causes of the explosion were systemic and related to communication failures. "Most, if not all, of the failures at Macondo can be traced back to underlying failures of management and *communication*. Better . . . *communication* within and between BP and its contractors . . . would have prevented the Macondo incident. . . . Information appears to have been excessively departmentalized at Macondo as a result of poor communication. . . . As a result, individuals often found themselves making critical decisions without a full appreciation of the context in which they were being made (or even without recognition that the decisions were *critical*" (emphasis added).[29]

Problems in Communicating with Electronic Media

The problems discussed in the report to the president on the BP/*Horizon* disaster are representative of problems that can occur when electronic media are used to communicate complex issues. Distinct differences exist between texting and e-mailing and other, older, forms of communication, such as the telephone or traditional letters and memoranda. When writers blur these differences or fail to recognize them, messages may become confusing or unintelligible, and readers may misinterpret or fail to heed the messages at all.

There are four major problems inherent in electronic communication. These appear to be caused by the tendency of writers (1) to transfer the conventions and style of social media to technical and scientific correspondence, (2) to transfer the functions of the telephone to e-mail and texting, (3) to rely solely on electronic media to discuss complex issues or issues requiring immediate response, and (4) to eliminate reflection and revision. As a result, writers may omit information necessary for the readers' understanding, provide inaccurate information, include irrelevant information, and misplace information within a text.

1. Transfer of the Conventions and Style of Social Media to Technical/Scientific Correspondence

Although the majority of electronic correspondence related to technical/scientific information assumes the relatively formal, impersonal, and objective style traditionally followed in letters and memoranda, there is a tendency for writers to slip into the conventions and style of social media (e-mail, texting, Facebook, Twitter, LinkedIn, etc.), which occur in a high-context environment, are personal in focus, and use an informal, conversational style.

High Context. High context relates to cultural groups in which individuals share similar experiences and knowledge. A high-context message omits background information and other data that the writer assumes the reader knows.[30] Because such social media sites as Facebook and LinkedIn are sent to a closed group of people, all of whom have certain knowledge in common, communication within these sites involves high-context messages that assume readers are knowledgeable in the topics under discussion and are aware of the topics' background.

A problem in communication occurs when writers mistakenly believe that their audience—employees, managers, clients, and sub-contractors in large corporations and organizations—are similar to their readers on LinkedIn and Facebook and write to them as if they were in a high-context environment. In technical/scientific business environments, readers may be located in different geographic locations, have different specialties, and be assigned to different divisions. They often need background information and elaborated details to under-stand a message. When writers fail to realize that they are writing in a low- rather than a high-context environment, the information their readers need may be omitted.

Several days after the explosion, one of the engineers who had been aboard the rig e-mailed BP vice president of drilling and comple-tions Pat O'Bryan with the bladder explanation (Figure 4.5). O'Bryan responded with a series of question marks—that is, the explanation didn't make sense.

It has been suggested that if the bladder effect explanation had been communicated to O'Bryan when it was first discussed on the rig, he would have invalidated it and halted the tests. Perhaps. But the rhe-torical strategies used in the message below in Figure 4.5 do not lead me to believe that such would have been the case. Both messages are writer based. Mike fails to overtly state his request—that is, if you do not agree with this rationale, please tell me. Pat's response is simply

I believe there is a bladder effect on the mud below an annular preventer as we discussed. As we know the pressure differential was approximately 1400-1500 psi across an 18 'A "rubber annular preventer. 14.0 SOBMplus 16.0 ppg [pounds per gallon] Spacer in the riser, seawater and SOBM below the annular bladder. Due to a bladder effect, pressure can and will build below the annular bladder due to the differential pressure but cannot flow - the bladder prevents flow, but we see differrential pressure on the other side of the bladder.
Now consider this. The bladder effect is pushing 1400~1500^>siagaiast-allj^.ta&jnudbeJowrMe, have displaced to seawater from 8,367'to just below the annular bladder where we expect to have a 2,350 psi negative pressure differential pressure due to a bladder effect we may only have a 850-950 psi negative pressure until we lighten the load in the riser.
When we displaced the riser to seawater, then we truly had a 2,350psi differential and negative pressure.

Mike,

???
Regards,
Pat

FIGURE 4.5. *Example of an ambiguous e-mail in which the writer assumes the reader understands his message*

a series of question marks, implying that Mike will know what the question marks signify. However, his questioning could mean any of the following: (1) I don't understand the concept being discussed and would like further explanation; (2) I don't know why are you sending this explanation to me; or (3) I don't know where in the world this explanation comes from, but it doesn't make sense.

The communication threads in Figure 4.6 are an example of a successful chain in which participants provide each other with background information and sufficient detail to understand a message. The chain, concerned with pip tags and casings, involves three persons: Sarah Dobbs, who is with onshore BP engineering; Brian Morel, also an onshore BP engineer but in a different division; and Mark Hafle, another onshore BP senior drilling engineer. Although Hafle is included as a recipient of the message, he does not become involved in the conversation.

Dobbs is very aware of her readers. In the first communication, she includes background "historical knowledge" that she feels her readers need but may not know. The threads in the chain are coherent. Each participant responds directly to the previous message through topic coherence (repetition of the key word in the previous message). In the first message, Dobbs asks why the decision was made to use seven-inch liners, and the first line in Morel's message reiterates the data, seven

From: Dobbs, Sarah
Sent: Tuesday, March 30,2010 10:33 AM
To: Morel, Brian P; Hafle, Mark E
Cc: PINEDA, FRANCISCO
Subject: Pip Tags and Casing

Guys -
We would like pip tags near the casing hanger, just below the
production liner top, just above the sand, and one near TD.
Also, what swayed the decision to 7" liner?   Was it availability or cementing concerns?
And, for historical knowledge, in August it looks like y'all discussed
(with Hu) the option of 10-3/4" casing to 3000' below the mudline to
accommodate a larger 1 5K SCSSV.   It was our understanding that
the casing was unavailable, so that option was eliminated. Is that
correct?
By the way, do y'all know if there is an SoR floating around for this well.

Sarah Elisabeth Dobbs
BP Gulf of Mexico Deepwater
Completions Engineer

From: Morel, Brian P
Sent: Tuesday, March 30,2010 10:54 AM
To: Dobbs. Sarah; Hafle, Murk E
Cc: PINEDA, FRANCISCO
Subject: RE. Pip Tags and Casing

7" is so we can run a long string instead of a tieback and still cement If we had run
7-5/8" we would not have been able to cement as a long string with the amount of
casing available, and would have had to double Hie source -3000' extra to bring ilie
7-5/8" above Hie 11-7/8" lianger a few hundred feel.   We did not have this pipe in
stock or easily available, and were able to get T from Ncxcn.

Not running the tieback, saves a good deal of time/money as well as reduces
complexity due to the conventional casing hanger or having to run a second packer
and PBR assembly for the Versaflex hanger. If you want more details let me know.

We will take care of getting the pip tags. Do you want the casing hanger one moved
to the top of the cross-over instead?
Thanks
Brian

From: Dobbs, Sarah
Sent: Wed Mar 31 21:15:192010
To: Morel, Brian P
Subject: RE: Pip Tags and Casing
Importance: Normal

Yes, that is great. Thanks.

Sarah Elisabeth Dobbs
BP Gulf of Mexico Deepwater
Completions Engineer

FIGURE 4.6. *Example of an effective e-mail chain using topic coherence*

inches, ensuring Dobbs can connect her question to his response even if she has been involved in other matters during the twenty minutes between messages and has forgotten the point of her previous e-mail. Morel provides a detailed explanation for the decision rather than a single-word answer. His final paragraph responds to the request Dobbs makes in her first paragraph. Finally he asks her a question to which she responds in the third message in the chain.

Personal Focus. (Con)fusing the style of social media with that of business/technical messages, writers may interject the "I" aspect of social media into electronic technical/scientific communication, creating a writer- rather than a reader-based message. This (con)fusion often leads writers to include personal information not related to the focus of the message and of no interest to readers, resulting in a TMI response (Too Much Information) as occurs in Figure 4.4.

In Figure 4.7, an e-mail chain three days before the blowout took place between Houston-based BP wells leader John Guide and his manager, David Sims, BP's onshore manager for Gulf of Mexico oil drilling operations, also in Houston. Guide, who sends the e-mail as a result of a revision in the cementing procedures of the well, vents his frustration with the changes that he perceives constantly occurring on the rig. Sims's response sounds more like the response to a social media text message than to a call for help. The comment about dancing is not only irrelevant to the topic but irreverent in light of the consequences that followed.

Informality. By transferring the informal style of social media to technical or scientific correspondence, writers may lapse into either a staccato syntactic style more consistent with a tweet or a rambling style more consistent with a personal blog. The tendency to respond monosyllabically or with minimal wordage (the 140 characters of a tweet) can leave readers with incomplete responses to a discussion topic, as in Figure 4.5. Responders either may need to continue the chain by requesting further information or may leave the chain with incomplete information that could result in misunderstanding a message or the inability to follow through or make a decision. On the other hand, the tendency to ramble increases the wordage in a message and may delay the presentation of the main point, creating a text that may

David, over the past four days there has[sic] been so many last minute changes to the operation that the WSL 's [well site leaders] have finally come to their wits end The quote is —flying by the seat of our pants. Moreover, we have made a special boat or helicopter run every day.

Everybody wants to do the right thing, but this huge level of paranoia from engineering leadership is driving chaos. This operation is not Thunderhorse [one of the world's largest rigs ever built. The rig was damaged by Hurricane Dennis.] Brian has called me numerous times to make sense of all the insanity. Last night's emergency evolved around 30 bbls [barrels] of cement spacer behind the topping and how it would affect any bond logging (1 do not agree with putting the spacer above the plug to begin with). This morning Brian called me and asked my advice about exploring other opportunities both inside and outside of the company.

John, I've got to go to dance practice in a few minutes- Let's talk this afternoon. for now, and until this well is over, we have to try to remain positive and remember what you said below - everybody wants to do the right thing. The WSLs will take their cue from you. If you tell them to hang in there and we appreciate them working through this with its (12 hours a day for 14 days) - they will. It should be obvious to all that we could not plan ahead for the well conditions we 're seeing, so WE have to accept some level of last minute changes.

We 've both [been] in Brian's position before. The same goes for him. We need to remind him that this is a great learning opportunity, it will be over soon, and that the same issues - or worse - exist anywhere else. I don't think anything has changed with respect to engineering and operations.

Mark and Brian write the program based on discussion "direction from you and our best engineering practices. If we had more time to plan this casing job, I think all this would have been worked out before it got to the rig. If you don't agree with something engineering related, and you and Gregg can't come to an agreement, Jon or me gets involved. If IT's purely operational, it's your call.

I'll be back soon and we can talk,
We're dancing to the Village People!

FIGURE 4.7. *Example of an e-mail (con)fusing social media and business communication*

cause readers to stop reading, which is probably what occurred when the NASA team received the letter in Figure 4.4.

Writers following the style of social media may also substitute the informal conventions of social media for the more formal ones of business. Although her e-mail is successful overall, Dobbs in Figure 4.6 uses the informal southern pronoun "y'all." Rather than using the more formal salutation of "Good Morning" or simply the name of the reader, e-mails that conflate the social style with a business message may begin with the social media salutation "Hi," which may not be appropriate for addressing a specific reader, especially one who comes from an Asian or South American culture. In Figure 4.6, Dobbs's salutation "Guys" is not only informal, but, if adapted to women, "Ladies" or "Gals" could be (mis)interpreted as sexist. In addition, writers may employ the shorthand spelling and idioms of the electronic media, such as TMI, with which some readers may not be familiar.

2. TRANSFER OF THE FUNCTIONS OF THE TELEPHONE TO E-MAIL AND TEXT MESSAGES

The confluence of the functions of the telephone with e-mail and text messages has resulted in a misperception that communication is two-way and that conversations occur simultaneously and in real time.[31] However, in most cases they are asynchronous.

During a telephone conversation, a person can question an aspect of a description, ask for the definition of a term, or request a fuller explanation of a topic and receive an immediate response, even if it is "I don't know" or "I'll get back to you." However, during an e-mail or text conversation, the communication can be suspended before a response is made. Of the approximately half of all messages that people read, only about one-third of these ever receive a response.[32] Yet writers not only want a response but want it immediately. This is seldom the case, with the average time for a response being approximately twenty-eight hours.[33]

The gaps in time between messages in an e-mail chain are caused by at least one of the participants leaving the conversation before it is concluded, often because the participant has turned to other work. Whether or not a person responds to a message immediately is usually determined by his or her perceptions of the importance of a message. Other criteria include the amount of time required for writing a response. Messages requiring more than a quick response are often put aside.[34]

New conventions have been introduced to alleviate the pressure for readers to respond immediately to every message as soon as it arrives.

It has become acceptable practice to notify the sender that the message has arrived without providing a response to the specific topic in the message. The first message in the chain in Figure 4.8, which is concerned with the need to obtain a liner, occurs over a three-hour period, between 9:39 AM and 12:26 PM. The writer of the first message in the chain is aware of the readers' felt need to respond immediately to his message and makes an explicit exception to this convention. Recognizing the need for a gap between messages to allow the recipient time to obtain the information he is requesting, he uses all caps to ensure the readers read the injunction that he does not want a response until the recipients can provide the information he needs. There is about an hour gap between the threads of this chain.

Because of the time gap between e-mails and texts in a chain, writers need to provide coherence not only within a message but between message threads. To remind the participants in a chain of the question or request that was made in the previous thread, the first line of each message should explicitly refer to the topic, request, or question of the previous message. Such a reminder is especially helpful if several messages have intervened between the initial request and the response or if several participants are involved in the chain or several topics are under discussion. Figure 4.9, a chain involving three people, takes place over a ten-hour period. It is an example of an e-mail chain in which the participants provide prompts to remind each other of the topics under discussion. Hafle notifies the other two participants that he has read their messages by thanking Miller and paraphrasing Morel's comment "This has been a nightmare well" with the comment "This has been a crazy well for sure." However, despite the problems on the rig, Miller cuts the thread, indicating he will be going out of town. Yet, both Sims's and Miller's messages relate to issues that will be identified later as key to the blowout.

3. Sole Reliance on Electronic Media to Discuss Complex Messages

Often, because of the ease of sending electronic messages, other forms of media are rejected. Yet, if issues are complex or open to

From: Couch, Thomas E

Sent: Wednesday, March 24,2010 9:39 AM
To: Smith Ridley; Alfonso de las Cuevas; Crane, Allison; | •IHBHH0@cabotog.com; ••BB@petrohunt.coin;
Subject: BP is looking For 7" Liner

STA,
BP is looking for approx 5,500 feet of 7.00", 32#, Q-125, Hyd. 513 or 523 Liner. Will look at
other connections and possibly PI 10 material
PLEASE DO NOT RESPOND UNLESS YOU HAVE SOMETHING TO OFFER.

Regards and Thanks,
Tom Couch
BP Exploration & Production Co.
PSCM, Materials Management, GoM
SPU

From: Rincon, Pal (Dallas) *i*
Sent: Wednesday, March 24,2010 10:52 AM
To: Couch, Thomas E
Subject: RE: BP is looking for 7" Liner
Tom,

Nexen has 6,000' - *T* 32# Q-125 HC Hydril 513 Condition A -
recently inspected

What's your need date?

Subject to management approval - will pursue if you are interested.

Pat C. Rincon
Sr Buyer, Supply Mgrat.

Nona Petroleum US A. Ine

From: Couch, Thomas E
Sent: Wednesday. March 24.2010 12:26 PM
To: Rincon, Pat (Dallas)
Cc: Crane, Allison
Subject: RE: BP is looking for 7" Liner
Pat,
This looks really good. Still don't have exactly what they're going to do in this well yet.

If it's not too much trouble do you have the following that you could send?
1. The location of the pipe
2. The recent inspection report(s)- we may not need to reinspect.
3. The Mill or MTR(s) even better.
Just as soon as we know the outcome of the requirement, we'll let you know.

Regards and Thanks,
Tom Couch
BP Exploration & Production
Co. PSCM, Materials
Management, GoM SPU

FIGURE 4.8. *Example of an effective e-mail chain*

misunderstanding, multiple forms of communication, including the
telephone, letters/memoranda, fax, and person-to-person meetings
may need to be used.[35] Both the telephone and person-to-person

From: Morel, Brian P
Sent: Wednesday, April 14, 2010 1:31 PM
To: Miller, Richard A
Cc: Hafle, Mark E
Subject: Macondo APB

Rich,
There is a chance we could run a production liner on Macondo instead of the planned long string. As this does not change much for APB based on the original design assumptions of a trapped annular, I don't see any major effects, but wanted to confirm I am not missing something. Attached is the proposed schematic, please let me know if you have any questions. We could be running it in 2-3 days, so need a relative quick response. Sorry for the late notice, this has been nightmare well which has everyone all over the place.

Thanks
Brian

From: Miller, Richard A
Sent: Wednesday, April 14, 2010 6:03 PM
To: Morel, Brian P
Cc: Hafle, Mark E
Subject: RE: Macondo APB

We have flipped design parameters around to the point that I got nervous. I did a rough update of both my disk calculations and my WellCat model. All looks fine.

If we run the 9-7/8" x 7" as a long string, then the design *resembles* the original configuration, at least from an APB standpoint. The outward-acting 16" rupture disks mitigate 9-7/8" collapse loads due to B annulus APB. I do not have the final disk depth, so I guessed it is around 9,500'.

If the 9-7/8" x 7" is run as a liner (per your schematic), then there is a risk that a trapped annulus forms between the 7" and 9-7/8" liners. The WellCat model predicts an incremental 2,350 psi APB in that annulus. To keep the 7" from collapsing, the pressure inside the 7" at 17,157' TVD needs to be 4,800 psi or greater. Assuming that the production packer is set above this depth, then the 4,800 psi could dictate a reservoir abandonment pressure limit. We can hash this out in the completion phase, but you may want to alert completions of that possible issue.

Let me know if you have questions. I'll be in Westlake Thursday morning and have an early afternoon flight to catch.
Rich

From: Hafle, Mark E
Sent: Wed Apr 14 23:09:46 2010
To: Miller, Richard A
Subject: RE: Macondo APB
Importance: Normal

Thanks Rich. This has been a crazy well for sure.

Mark

FIGURE 4.9. *Example of a coherent chain that is suspended by the second participant, who probably won't respond to his message for over twenty-four hours*

meetings provide real-time one-on-one opportunities to discuss a topic or clarify a misunderstanding during a single time slot, not over a period of time.

Stack, president of an international consulting company in Denver, Colorado, claims that "communication becomes richer as you add human elements like voice, tone, facial expression and physical expression." She has developed the flowchart in Figure 4.10 ranging from person-to-person communication for ambiguous, long, or difficult messages to letters and reports for clear, simple messages. She places e-mail in the middle of the range.[36] Melcrum has developed a more complex model.[37]

In Figure 4.10, after reading an e-mail from Guide, Walz recognizes there is a problem with the centralizers and requests that Guide discuss the situation over the telephone, by which the two can get into more detail and can talk through the confusing situation. He then requests a telephone call with Sims. In this case, although the e-mail provides a satisfactory means of communicating the need for a meeting, it does not provide the appropriate platform for clarifying confusion in a situation.

4. Failure to Read, Reflect on, and Revise a Draft

Electronic media, while providing quick and easy transmission, are also truncating the revision process because of the ease with which the "enter" and "send" buttons can be accessed. As a result, messages are fraught with grammatical and spelling errors, such as those in Figure 4.7. Although these mistakes do not create a problem in communicating a message, they may provide an impression of carelessness and haste. Much has been written about the need to pause before hitting the "send" button if a message has been written in anger.[38] However, it is also necessary to pause when messages contain complex information or are responses to complex situations. The need to reread a text to ensure that all of the information necessary for the reader to understand a message has been included is well documented. Yet, e-mail and text messages are seldom reread. Once a message is written, the "enter" or "send" button is hit. It is doubtful that Sims would have responded as he did if he hadn't been in such a hurry to go dancing.

Conclusion

The problems faced by readers attempting to understand writer-based messages are exacerbated by the problems associated with electronic communication. In order to create texts that are effective in achieving their purposes, writers need to avoid the four pitfalls outlined.

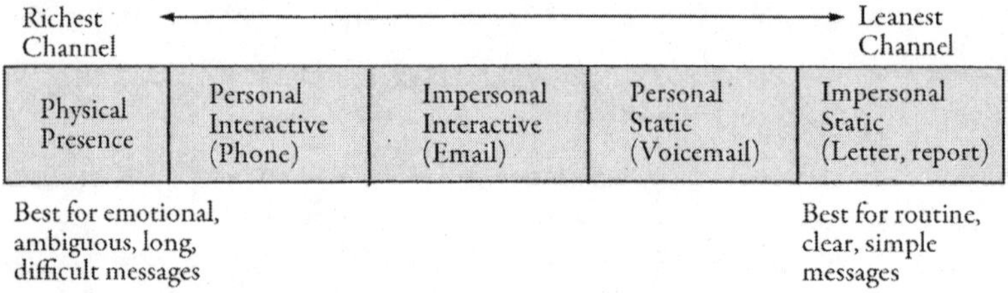

Richest Channel ←————————————————————————→ Leanest Channel

Physical Presence	Personal Interactive (Phone)	Impersonal Interactive (Email)	Personal Static (Voicemail)	Impersonal Static (Letter, report)

Best for emotional, ambiguous, long, difficult messages

Best for routine, clear, simple messages

FIGURE 4.10. *Stack's chart*

RECOMMENDATIONS FOR WRITING EFFECTIVE ELECTRONIC MESSAGES

Based on the present research related to electronic communication technology, the following recommendations provide ways to communicate effectively with e-mail and text messaging:

1. Keep messages brief, limited to a single screen if possible.
2. Use the inverted pyramid pattern, with the most important information first. KISS (Keep It Simple, Stupid) has always been a key acronym for writing business correspondence. For writing for electronic media, this approach is even more appropriate. The need to place the most important information at the beginning of a document becomes a necessity with electronic media.
3. Use the subject line to specify exactly what the message is about. Keep the subject line short, no more than forty characters.
4. Follow the conventions for business/technical correspondence:
   a. In a salutation use either the recipient's name or omit a salutation altogether.
   b. Provide a business-style closing followed by the name of the sender.
   c. Use a business or semiformal style.
5. Refrain from including personal or irrelevant information.
6. If a response is required, indicate to the recipient when the response is expected.
7. When responding in a chain of messages, always indicate the link to the previous message by referring in some way to the topic under discussion so that there is coherence between threads.

8. If you will not be responding immediately to a question or directive in a chain, respond with an estimate of when you will be able to continue the chain.

9. Respond to all messages to notify the sender that a message has been received. Even if a response is not required to answer a question or provide information, a message should be sent notifying the sender that the message was received.

10. Consider using other forms of communication:

    a. If you need to communicate complex information that a recipient may have difficulty understanding, consider using a telephone rather than e-mail or a text.

    b. If you need to communicate detailed explanations, consider attaching the information to an e-mail rather than including it on the e-mail or text message.

11. Use a sans serif font for electronic communication.

12. Remember the 3 R's: Reread, Reflect on the content and style, and Revise a message when necessary.

Notes

1. Karianne Skovholt, "Email Literacy in the Workplace" (Dissertation, University of Oslo, 1981), 1.

2. Sylvia Scribner and Michael Cole, *The Psychology of Literacy* (Cambridge: Harvard University Press, 2009); Skovholt, "Email Literacy in the Workplace," 236.

3. Lauren Ekroth, "Have Email Conversation Problems?" *DonMorris*, accessed February 2016, http://donmorris.com/article/have-email-conversation-problems; Kristin Byron, "Carrying Too Heavy a Load? The Communication and Miscommunication of Emotion by Email," *Academy of Management Review* 33, no. 2 (2008): 309–27.

4. Zoltán Gócza, "Myth #3: People Don't Scroll," *UX Myths*, 2012, http://uxmyths.com/post/654047943/myth-people-dont-scroll.

5. University of Bath, "Effective E-mail," 2016, http://www.bath.ac.uk/bucs/email/guidelines/effectiveemail.

6. Byron, "Carrying Too Heavy a Load?"

7. Laura A. Dabbish, Robert E. Kraut, Susan Fussell, et al., "Understanding Email Use: Predicting Action on a Message" (2005), http://www.cs.cmu

.edu/~kiesler/publications/2005pdfs/2005-Dabbish-CHI.pdf; L. Louhiala-Salminen, "From Business Correspondence to Message Exchange: What Is There Left?," in *Business English: Research into Practice*, ed. C. Nickerson and M. Hewings (New York: Longman, 1999), 100–114.

8. Kleiner, Perkins, Caulfield, and Byers, "KCPB Internet Trends," May 29, 2013, http://www.slideshare.net/kleinerperkins/kpcb-internet-trends -2013.

9. Marketing Charts Staff, "When Smartphone Users Check Email during the Day," *American Writers and Artists Inc.*, May 2014, http://www .marketingcharts.com/online/when-smartphone-users-check-email-during -the-day-41401/.

10. John Wood, "The Best Fonts to Use in Print, Online and Email," *American Writers and Artists Inc.*, October 2011, http://www.awaionline .com/2011/10/the-best-fonts-to-use-in-print-online-and-email/; A. Dawn Shaikh, "The Effects of Line Length on Reading Online News," July 5, 2005, http://psychology.wichita.edu/surl/usabilitynews/72/linelength.asp; B. S. Chaparro, A. D. Shaikh, and A. Chaparro, "Examining the Legibility of Two New ClearType Fonts," *Usability News*, February 2006, http://usabilitynews .org/examining-the-legibility-of-two-new-cleartype-fonts/.

11. Jakob Nielsen, "F-Shaped Pattern for Reading Web Content," *Nielsen Norman Group*, April 2006, http://www.nngroup.com/articles/f-shaped -pattern-reading-web-content/; Jakob Nielsen, "How Users Read on the Web," *Nielsen Norman Group*, October 1997, http://www.nngroup.com /articles/how-users-read-on-the-web/; Steve Outing, "Eyetrack III: What News Websites Look Like through Readers' Eyes," *Poynter*, March 2011, http://www.poynter.org/uncategorized/24963/eyetrack-iii-what-news -websites-look-like-through-readers-eyes/; National Dissemination Center for Children with Disabilities, "How People Read on the Web," *Center for Parent Information and Resources*, August 2012, http://www.parentcenter hub.org/repository/web-reading; Sara Dickenson Quinn, "New Poynter Eyetrack Research Reveals How People Read News on Tablets," *Poynter*, October 2012, http://www.poynter.org/how-tos/newsgathering-storytelling /visual-voice/191875/new-poynter-eyetrack-research-reveals-how-people -read-news-on-tablets/.

12. Anujeet Mujumdar, "Smartphone Users Check Their Phones an Aver-age of 150 Times a Day," *Tech2*, May 30, 2013, http://tech.firstpost.com/news -analysis/smartphone-users-check-their-pnones-an-average-of-150-times-a -day-86984.html.

13. Radicati Group, "Email Statistics Report, 2011–2015," May 2011, http://www.radicati.com/wp/wp-content/uploads/2011/05/Email -Statistics-Report-2011-2015-Executive-Summary.pdf.

14. Thomas Jackson, "Email Stress," 2012, http://profjackson.com/email _stress.html; Stephen R. Barley, Deborah E. Myerson, and Stine Grodal, "Email as a Source and Symbol of Stress," August 2011, http://people.bu.edu/grodal /Email.pdf; Daantje Derks and Arnold B. Bakker, "The Impact of E-mail Communication on Organizational Life, *Cyberpsychology (Brno)*, May 2013, http:// www.cyberpsychology.eu/view.php?cisloclanku=2010052401&article=4.

15. Skovholt, "Email Literacy in the Workplace."

16. Information in this section is from Columbia Accident Investigation Board, *Report of the Columbia Accident Investigation Board*, vol. 1 (Washington, DC: Government Printing Office, 2003), 1.

17. Ibid., 140.

18. Skovholt, "Email Literacy in the Workplace."

19. Nielsen, "F-Shaped Pattern for Reading Web Content."

20. For general information, see National Commission on the BP Deepwater Horizon Oil Spill and Offshore Drilling, *Deepwater: The Gulf Oil Disaster and the Future of Offshore Drilling; Report to the President* (Washington, DC: US Government Printing Office, 2011); Joel Achenbach, *A Hole at the Bottom of the Sea: The Race to Kill the BP Oil Gusher* (New York: Simon and Schuster, 2011); Andrew Hopkins, *Failure to Learn: The BP Texas City Refinery Disaster* (Sydney: CCH Australia Ltd., 2010); Andrew Hopkins, *Disastrous Decisions: The Human and Organisational Causes of the Gulf of Mexico Blowout* (Sydney: CCH Australia Ltd., 2012); John Konrad and Tom Shroder, *Fire on the Horizon: The Untold Story of the Gulf Oil Disaster* (New York: HarperCollins, 2011).

21. Darryl Fears, "Deepwater Horizon Oil Left Tuna, Other Species with Heart Defects Likely to Prove Fatal," *Washington Post*, March 24, 2014, http://www.washingtonpost.com/national/health-science/after-deepwater -oil-spill-once-speedy-tuna-no-longer-make the grade/2014/03/24/4d2c2d 78-b378-11e3-b899-20667de76985_story.html; National Academy of Sciences, *Proceedings of the National Academy of Sciences*, 111, no. 32 (2014): 128, 193–194, 199–202.

22. Scott Bronstein and Wayne Drash, "Rig Survivors: BP Ordered Shortcut on Day of Blast," *CNN*, June 9, 2011, http://www.cnn.com/2010/US/06 /08/oil.rig.warning.signs/.

23. Bartlit, Beck, Herman, Palenchar & Scott, LLP, "Presidential Oil Spill Commission Report from Chief Council, Fred Bartlit," February 2011.

24. Bronstein and Drash, "Rig Survivors."

25. National Commission, *Deepwater*, 94.

26. National Commission, *Deepwater*, 25.

27. Associated Press, "Gulf Oil Spill Deaths: The 11 Rig Workers Who Died during the BP Deepwater Horizon Explosion," *Huffington Post*, November 15, 2012, http://www.huffingtonpost.com/2012/11/15/gulf_oil _spill-deaths_n_2139669.html.

28. Michael Kunzelman, "Rig Supervisors Robert Kaluza and Donald Vidrine Challenge Manslaughter Counts," *Huffington Post*, May 13, 2013, http://www.huffingtonpost.com/2013/05/31/bp-rig-supervisors-robert -kaluza-donald-vidrine_n_3366355.html.

29. Quote from National Commission, *Deepwater*, 122–23. Ripley suggests that this is also a psychological phenomenon, reflecting a "normalcy bias" in which people, especially those in charge of a project, resist accepting that situations are not as they want them to be. Amanda Ripley, *The Unthinkable: Who Survives When Disaster Strikes—and Why* (New York: Three Rivers Press, 2008).

30. Edward Hall, *Beyond Culture* (New York: Anchor Books, 1976).

31. Simon Scerri, "Aiding the Workflow of Email Conversations by Enhancing Email with Semantics" (Galway, Ireland: Digital Enterprise Research Institute, 2007), http://aran.library.nuigalway.ie/xmlui/handle /10379/500.

32. Dabbish et al., "Understanding Email Use"; Skovholt, "Email Literacy in the Workplace."

33. Farshad Kooti, Luca Maria Aiello, Mihajito Grbovic, et al., "Evolution of Conversations in the Age of Email Overload," Proceedings of the 24th International Conference World Wide Web, Florence, Italy (2015): 603–13, http://www-scf.usc.edu/~kooti/files/kooti_email.pdf.

34. Janelle Estes, "Email Subject Lines: 5 Tips to Attract Readers," *eNewsline*, May 4, 2014, http://www.nngroup.com/articles/email-subject -lines/.

35. Tanu Ghosh, JoAnne Yates, and Wanda Orlikowski, "Using Communication Norms for Coordination: Evidence from a Distributed Team," *Research Brief*, October 2004, http://aisel.aisnet.org/icis2004/10; Sally O. Hastings and Holly J. Payne, "Expressions of Dissent in Email: Qualitative Insights into Uses and Meanings of Organizational Dissent," *Journal of Business Communication* 50, no. 3 (2013), http://dx.doi.org/10.1177 /0021943613487071; Virginia Kupritz, "Productive Management Communication: Online and Face-to-Face," *International Journal of Business*

Communication 48, no. 1 (January 2011), http://job.sagepub.com/content /48/1/54.abstract; Anna K. Turnage and Alan K. Goodboy, "E-mail and Face-to-Face Organizational Dissent as a Function of Leader-Member Exchange Status," April 2014, http://job.sagepub.com/content/early/2014/o 4/13/2329488414525456.full.pdf.

36. Laura Stack, "Choose the Most Productive Communication Channel for Your Message," http://office.microsoft.com/en-us/help/choose-the-most -productive-communication-channel-for-your-message-HA001194524.aspx (no longer available).

37. Melcrum, "Choosing the Right Communication Channel," https:// www.melcrum.com/research/strategy-planning-tactics-intranets-digital| -social-media/choosing-right-communication (acquired 2015).

38. Stuart Chalmers, "Dang, I Hit the Send Button Too Soon," accessed February 2016, http://digitalbloggers.com/normalguy/dang-i-hit-the-send -button-too-soon/; Miss Manners, "Stop, Reread and Think before You Hit Send," 2012, http://www.businessemailetiquette.com/stop-reread-and-think -before-you-send/.

5 Communicating with PowerPoint

The Army, NASA, and the Enbridge Pipeline

The advent of electronic media has not only made it easier and quicker to communicate text, but also has made it easier to provide graphic representations. The ease with which graphics and text can now be interrelated is evident in the ubiquitous use of slideware in oral presentations. However, as with the introduction of electronics in correspondence, the introduction of slideware—often incorrectly referred to as PowerPoint, which is only one of a number of software programs used for presentations—has been both a help and a hindrance in terms of an audience's ability to comprehend a message. In many cases, PowerPoint has been blamed for an audience's failure to learn from a message or to understand a text.

PowerPoint and the Army

One of the most (in)famous PowerPoint slides is a chart prepared by the military to illustrate the complexity of American intervention in Kabul (Figure 5.1) when the Afghanistan war was at its height. The slide was part of a presentation to senior military officers analyzing

DOI: 10.5876/9781607324676.c005

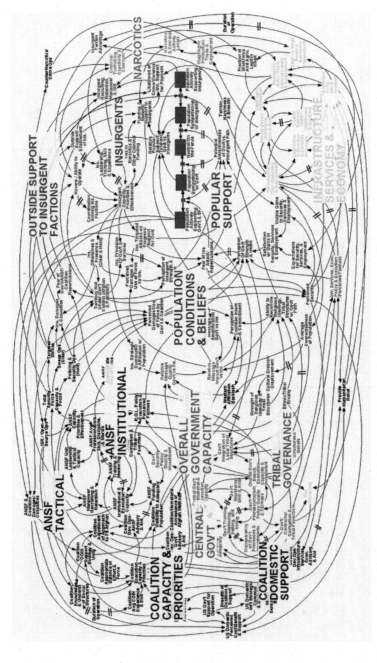

FIGURE 5.1. *The Afghanistan war chart slide* (New York Times, April 26, 2010)

the dynamics and challenges of the counterinsurgency (COIN) in Afghanistan.

After seeing the chart, General Stanley McChrystal is said to have commented, "When we understand that slide, we'll have won the war." According to the *New York Times* (*NYT*), which reproduced the chart under the headline "We Have Met the Enemy and He Is PowerPoint," a number of military commanders note that "some problems in the world are not bullet-izable . . . [PowerPoint] create[s] the illusion of understanding and control . . . [It] stifles discussion, critical thinking and thoughtful decision-making." Brigadier General H. R. McMaster banned PowerPoint presentations in 2005 when he led the effort to secure the Iraqi city of Tal Afar. Dr. Thomas X. Hammes, a retired marine colonel, suggests that giving a PowerPoint presentation is like "hypnotizing chickens,"[1] thus supporting Marshall McLuhan's 1967 prescient prediction "The medium is the message."[2]

But is McLuhan's conclusion valid? Is the medium the message or is it that writers have allowed the technology to control the message? The truth of the matter is that GIGO (garbage in, garbage out) applies: what comes out of the computer is only as good as what goes in. The problem is not PowerPoint, but the rhetorical and graphical decisions that are made. In too many cases, PowerPoint is writer/designer-based rather than reader/audience-based.[3]

For all its hype, PowerPoint is simply a traditional slide or overhead presentation in which a computer rather than a projector is used to display information. Many of the rules that were applicable for designing a slide presentation are just as relevant for designing a PowerPoint presentation: using typeface large enough for the audience in the back of the room to read; providing only a main idea, not an entire paragraph; refraining from putting too much information on a single slide; and avoiding visuals that cover words or detract from the audience's attention to the message.

Most people who create PowerPoints are familiar with these visual rules. The problems with PowerPoint relate to rhetorical strategies, those that underlie the contradictory complaints of Generals McMaster and McChrystal—that such presentations fail to show the complexity of a problem or that they are so complex that they are unintelligible.

These are the same errors as those found in technical/scientific texts. The writers/designers have failed to take into account their audience/readers and, as a result, have made inappropriate rhetorical decisions.

Recommendations to improve presentations include adapting new software presentation programs,[4] using different template formats,[5] changing the syntactic structure of a presentation by using full sentences rather than elliptical ones,[6] and eliminating bullets.[7] In addition, presentations often suffer from the same weakness as computer documentation: they provide a canvas for designers to show off their bells and whistles rather than provide users with the information they need to engage in their tasks or to make decisions.

However, these recommendations fail to deal with the underlying root causes of the problems, which are twofold: (1) inappropriate rhetorical and graphical decisions that are writer/designer-based rather than audience/reader-based and (2) the synergistic effects of the medium and its ancillary parts, including handouts and note-taking. These ancillary parts are an integral part of a presentation and, when added to listening to a speaker and reading text on a screen, overload an audience's cognitive capacity, induce passivity, and reduce the audience's ability to understand the information.

This chapter looks at the presentation problems blamed on PowerPoint and the root causes of these failures in communication. It also provides suggestions for controlling the technology to provide effective presentations. Because PowerPoint is only one of a number of software programs that enable designers to provide a visual accompaniment to their oral presentations, the term "slideware" will be used for the remainder of this chapter, except for those instances in which PowerPoint is the slideware under discussion.

Inappropriate Rhetorical Decisions

Problems in communication blamed on slideware parallel those often found in engineers' traditionally written texts: inappropriate focus, organizational patterns, or sequence of information; omission of necessary details; and failure to consider an audience's needs. Writers' failure to make appropriate rhetorical decisions, regardless of whether they are writing a script for a slideware presentation or the text of a written document, results more often than not in miscommunication. This failure was the cause of the problem with the notorious slide in the PowerPoint presentation made during the *Columbia* shuttle disaster that has been blamed by many for the failure of NASA to attempt a rescue mission.[8] But blaming PowerPoint for the death of the crew is rather like shooting the messenger (in this case the slideware program) for the message.

COLUMBIA SHUTTLE BREAKUP

When the *Columbia* shuttle blasted off into space from Cape Kennedy in Florida on January 16, 2003, one of the pieces of foam used to protect the shuttle from the heat on reentry fell off, striking the shuttle before floating into space. Over the next sixteen days, the *Columbia* crew and Mission Control frantically tried to determine the damage that had been caused by the loose foam in an effort to determine whether or not this damage would cause a problem on reentry and, if so, to find a solution for the problem. They were unable to determine the extent or location of the damage in time for the crew to return home.[9] (For a more detailed explanation of the shuttle breakup, see chapter 4.)

No one either on board the *Columbia* or at Mission Control was aware that the foam had fallen off until the day after the launch, when NASA's Intercenter Photo Working Group at the Marshall Space Center began reviewing photographs from the launch. The engineers, charged with determining whether the foam's impact had caused damage to the shuttle, needed to know the precise location where the foam had hit the shuttle and the exact size of the hole or dent, but none of the photographs that had been taken of the launch showed the actual impact or the damaged area after impact.

In an attempt to obtain some answers, the team requested that the Department of Defense send a plane to photograph the shuttle. The request was denied by NASA managers. Instead of a photograph, the team was told to use a theoretical model, known as "the Crater," and numerical analysis to determine the extent of possible damage.

However, the team argued that the model was not appropriate; it would not be valid for a large impact area. According to the Columbia Accident Investigation Board (CAIB) report, the Crater "uses a specially developed algorithm to predict the depth of a Thermal Protection System tile to which debris will penetrate. This algorithm [is] suitable for estimating small (on the order of three cubic inches) debris impacts."[10] The actual impacted area, according to the CAIB, turned out to be 400 times this size. This discrepancy was exacerbated by previous tests that had indicated that the algorithm predicted more severe damage than was observed, leading engineers to classify the Crater as a "conservative" tool—one that predicts more damage than has actually occurred. One final problem with the model was that it had never been used while a mission was in orbit.[11] These problems invalidated the model's results.

Despite these problems, Mission Control insisted the model be used and that a report on the results of the model's analysis be presented the following day. Engineers conducted the analysis and put together a report, which was presented to NASA's Management Evaluation Team. The PowerPoint presentation included the slide in Figure 5.2.

The PowerPoint presentation reinforced the belief by the management team that the impact was minimal and photographic evidence was never provided. The slide has been blamed as a major reason for the team's lackadaisical response to the presentation.[12] In its criticism of the slide, the CAIB comments that "it is easy to understand how a senior manager might read the PowerPoint slide and not realize that it addresses a life-threatening situation."[13]

The problem with the presentation resides with the rhetorical decisions the writer made. These decisions were writer-based rather than audience-based. PowerPoint simply accentuated the inappropriate decisions.

Although the Crater model wasn't appropriate, thus invalidating the results, the focus of the presentation is on the invalid results. The statement of the model's lack of validity, "Flight condition is significantly outside of test database," is placed at the end of a long list of other bulleted items in a secondary position with the details supporting the statement "Volume of ramp is 1920 cu in vs 3 cu in for test," placed even further down at a tertiary level. Further reducing the impact of

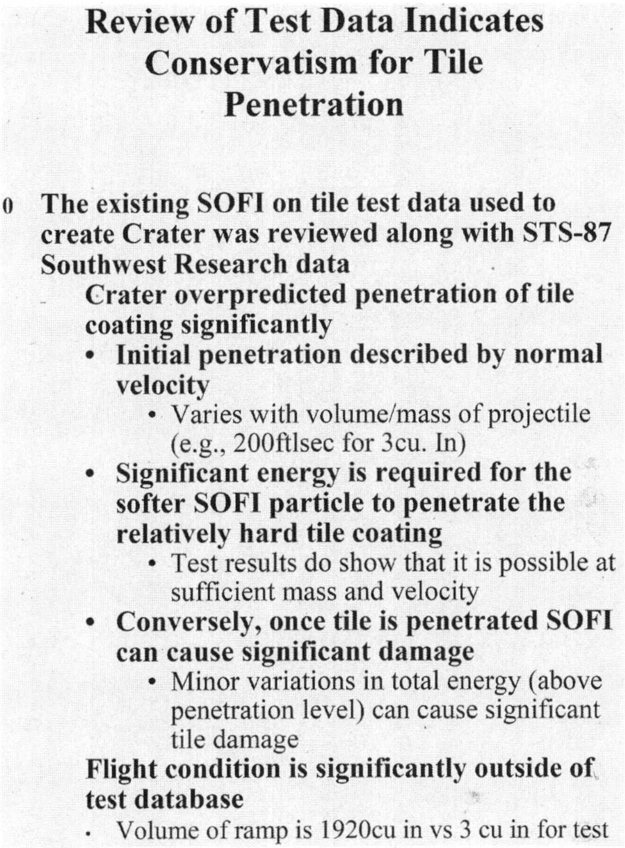

Review of Test Data Indicates Conservatism for Tile Penetration

0 **The existing SOFI on tile test data used to create Crater was reviewed along with STS-87 Southwest Research data**
- Crater overpredicted penetration of tile coating significantly
   • **Initial penetration described by normal velocity**
      • Varies with volume/mass of projectile (e.g., 200ftlsec for 3cu. In)
   • **Significant energy is required for the softer SOFI particle to penetrate the relatively hard tile coating**
      • Test results do show that it is possible at sufficient mass and velocity
   • **Conversely, once tile is penetrated SOFI can cause significant damage**
      • Minor variations in total energy (above penetration level) can cause significant tile damage
   Flight condition is significantly outside of test database
   · Volume of ramp is 1920cu in vs 3 cu in for test

FIGURE 5.2. *The infamous* Columbia *slide*

the model's lack of validity is the placement of the information indicating the model's tendency to overestimate: "Crater overpredicted penetration of tile coating significantly" is toward the top of the slide, providing readers with the impression that the data being presented could make the situation appear worse than it actually was.

The Columbia Accident Investigation Board has described the NASA culture at that time as insular, an environment in which employees' first concern was to protect themselves, reluctant to arrive at unpopular conclusions or make recommendations in opposition to administrators' goals. The managers had requested a presentation on the results of the test, having previously dismissed objections to its inadequacy, and therefore the writer organized his material in conjunction with these requirements.

The description of the NASA culture by the Investigation Board relates to the writer's decision to focus on the results of the model instead of on the model's inadequacy, thus subordinating the fact that the results are invalid. Once again cultural cognition plays a major role in the writer's decisions, resulting in an example of "garbage in, garbage out [GIGO]." (For additional information on cultural cognitions, see chapter 3, notes 1–4 in this book.)

Inappropriate Graphical Decisions

As with rhetorical decisions, inappropriate graphics, such as the one printed in the *New York Times*, are not caused by the slideware program but by designer-based decisions. Robert Browning, when asked to explain a certain poem, was said to have replied, "When I wrote this poem, only God and I knew what it meant. Now only God knows." The same could probably be said of the designer of the graphic viewed by General McChrystal. It is doubtful if even the designer of the *New York Times* graphic could explain it two weeks after he presented it.

A graph, photograph, schematic, or chart should be functional. As Michael Markel so deftly explains, "Functional graphics . . . help the audience understand . . . abstract concepts and relationships as well . . . Particular details . . . also help the audience understand the logic, organization, and development of the presentation."[14]

Synergistic Effects

The problems with slideware presentations are not limited to the slides. An oral presentation is multifaceted, including not only the verbal performance and individual slides, but also the slide deck that is the aggregate of all the slides, the accompanying handouts, and the notes taken by the audience. It is therefore necessary to study not only the individual aspects of a presentation but also the ways in which these entities interact and the interaction's synergistic effect upon the audience that, as Hammes notes, causes a "hypnotic effect." The interaction of the various elements results in cognitive overload, which has consequences in an audience's failure to learn and retain information as well as in audience members' inability to differentiate between significant and less important data.[15] It appears that when a verbal presentation is accompanied by slides that are combined into a full slide deck and when handouts of the slide deck are provided for the purpose of reference and note-taking, a synergistic effect occurs that creates cognitive stress and interferes with the learning process.[16]

DIFFERENTIATING EFFECTS BETWEEN
TEXTUAL AND GRAPHICAL SLIDES

In a PowerPoint presentation, when people try to listen to a speaker and simultaneously read the text on the screen, they use the same section of their brain, and this causes cognitive overload. However, when people view a graphic, picture, chart, or diagram while simultaneously listening to a speaker, they do not experience this problem.[17]

According to dual code theory, the brain processes visual and textual information in two different sections.[18] When a task requires that they be used together, as in viewing a picture and listening to a speaker, information is processed in each section without interference. However, when two tasks require a single section, as with reading the text of a slide and listening to a verbal presentation, especially if at least one of the tasks requires processing complex information, that section of the brain becomes overloaded.[19] This theory appears to explain some of the military's responses to PowerPoint slideware, which is often heavily text-based.

EFFECTS OF A SLIDE DECK

Members of an audience have difficulty discerning hierarchical differences when they view a full slide presentation, because the visual appearance of each slide in the deck is generally the same (sets of bullets or large masses of paragraphs).[20]

The format of the slide deck in a handout further impedes an audience's ability to take notes during a slideware presentation. Usually six slides are included on a page. This format, with the font often reduced to illegibility, not only creates difficulty in reading the text but also leaves little room for note-taking.

EFFECTS OF HANDOUTS AND NOTE-TAKING

When an audience is required to read a handout (again, using the language section of the brain) and synchronize it with the text on the screen, further overloading of the brain occurs.[21]

Handouts also indirectly affect an audience's ability to learn and retain information, because they give participants the perception that they do not need to take notes. Many members of an audience believe that handouts (usually a copy of the slide deck), disseminated either prior to or following a lecture, provide sufficient information to eliminate the need to take notes.[22] However, the writing-to-learn model assumes that through writing we learn. By taking notes, learners synthesize information and in so doing encode and chunk information for placement in long-term memory. When students or participants at a meeting fail to keep notes, much of the information is lost in the long term. Researchers have found that those who fail to take notes receive the lowest grades, those who read a summary provided by the lecturer receive the next lowest grades, and those who take notes receive the highest grades.[23]

Handouts not only appear to deter an audience from note-taking, but give an audience the impression that they do not need to listen attentively, as they will have the information to look at later. The audience watches and listens to the presentations in much the same way that they view a film or a television program, thus supporting McLuhan's "The medium is the message."

While using visuals does not usually interfere with listening to a speaker, note-taking does interfere because it requires the audience to encode in text the words they are hearing.[24]

Improving Presentations

If inappropriate rhetorical and graphical decisions and the synergistic effect of the various facets of a slideware program are the root causes of problems in learning and retaining information, then solutions need to address these aspects of a presentation.

CHANGING PERSPECTIVE

Perhaps the most needed change is that of perspective. Slideware should be perceived as *supporting* a message rather than *being* the message.[25] PowerPoint should be seen as a "beneficial communication technology that . . . enables presenters to provide appropriate visual support for their presentation. [It has the value of] persistently displaying the framework of the presenter's ideas to an audience."[26]

REDUCING SYNERGISTIC EFFECTS

If cognitive overload is caused by the simultaneous processing of three or even four tasks—listening to a lecture, reading text on a slide, reading text on a handout, and writing notes—then it would appear that either (1) these tasks need to be conducted separately so that they occur at different times, or (2) several of the tasks need to be eliminated or at least minimized. But there is conflicting evidence related to these solutions. Such ancillary procedures as note-taking and seeing an image have been found to improve learning and retention. In addition, members of an audience react differently to these various aspects.

The task of reading a slide's text appears to be a major factor in cognitive overload. However, because research indicates that many individuals learn better visually than aurally,[27] eliminating slides with text does not appear to be an acceptable solution. Research also indicates that people learn better if they are provided with a framework in which information is to be presented.[28] This would seem to support the use of providing an audience with visual text that presents at least the basic goals of a presentation and a simple outline of the major points to be covered. This step could take the form of either a slide or a handout to which a visual learner could refer and which an aural learner could ignore.

But again there's a rub. The use of handouts appears to be a cause of an audience's failure to learn as well as an audience's passive response to the information being presented. On the other hand, review of a presenter's notes seems to facilitate retention.[29]

The need for handouts could be eliminated if the audience took notes. As research has indicated the importance of note-taking for learning and retention, it would appear that this task needs to be retained. Because the parts of the brain used in writing and listening are in different locations, audiences have been able to engage in both of these tasks simultaneously. Thus, it would seem that presentations need to provide opportunities for the audience to take notes but that they need to be able to do so without the interference of either a textual or a visual factor.

The challenge is to find a way to present information that takes into account these conflicting interactions.

Guidelines

The following guidelines are offered as a way of presenting information with minimal interference among the various facets involved.

SLIDES

Slides with graphics

When presenters use slides with graphics, there is no cognitive overload as the audience is using two different sections of the brain—visual and linguistic—and the slideware is in a subordinate position. The following provide guidelines for effectively presenting graphic slides.

- Slides should be directly related to the topic under discussion.
- Slides should be used principally to present graphics that illustrate complex models and statistical information.
- Slides may contain text or stand alone as full-screen graphics, but text should be minimal.

Slides with textual content only

When presenters use slides with text, there is the potential for cognitive overload. To avoid this problem, such slides should be used minimally. The following guidelines apply to presenting text-only slides.

- Slides with text only should be limited to the following purposes.
  - To present a basic outline of the presentation that can be used at the beginning of a program to establish a framework for understanding the discussion. They can also prepare the listener for what is coming and be used at the end of the program to review the discussion. This format parallels the basic guideline of Toastmasters International: "Tell them what you're going to tell them, tell them, tell them what you've told them."
  - To present the major points of a topic and subtopic, providing for those members of the audience who are more visually than aurally oriented.
  - To provide quotes, statements, and points for emphasis that an audience should know verbatim.
- Slides with text should present a single concept, so that the audience has time to read it and to note it. The concept should be presented in sentence form to indicate the relationship of ideas. This is especially relevant if the audience needs to review the information later, since review of notes—whether the instructor's in the form of a handout or the student's—has been shown to be a factor in retention.
- When using slides with text only, speakers should pace their presentation according to the audience's ability to read the text. Audiences should be able to read a slide in ten seconds. If it takes longer than this, the slide needs to be shortened. By waiting to talk for ten seconds each time a new slide is shown, presenters should feel comfortable that their audience is ready to listen to their discussion without also trying to read the text on the screen and correlate it with their handouts.

Handouts and Note-Taking

Presenters can avoid adding to an audience's cognitive tasks by distributing handouts at appropriate times and by formatting handouts of slide decks to allow for sufficient note-taking.

- Handouts may be distributed in any of the following ways.
  - A representation of the entire slide deck may be distributed prior to the lecture.

- A handout providing only the basic outline of the presentation may be distributed prior to the presentation. The remaining handouts, whether they are duplications of the slide deck or other forms of presentation summary, may be disseminated after the presentation. This allows the audience to concentrate on the presentation and to take notes without the interference of handouts but provides for review of the information later.
- Handouts may be eliminated, allowing the audience to concentrate on the presentation and to take their own notes for review later.

- The format of a handout should provide sufficient space for listeners to take notes, which may include expanded explanations, examples of the information, and their own opinions of the ideas presented. No more than three slides should be included on a page that is divided into two columns, with the slides lined up vertically in the left column and space for notes in the right column. This format provides the audience with plenty of space to take notes related to each major topic or point discussed by the presenter.

By following these guidelines, the writer/designer gives preference to the verbal performance, with the need for the audience to read text minimized and the opportunity for the audience to take notes and to review the information at a later time maximized. In addition, by using fewer slides, a presenter's pace is slowed, thus giving note takers time to synthesize the information and express it in written form.

An Effective Slideware Presentation Mixing Graphics and Text: Expansion of the Enbridge Pipeline[30]

By 2012, the production of light crude oil in western Canada and North Dakota had increased exponentially. Enbridge Inc., an energy delivery company based in Canada, recognized that it needed to increase its capacity to transport oil from the North American oil fields to refineries throughout Canada and the United States. To move more oil, the company decided to replace portions of pipeline 6B, which ran between Griffith, Indiana, and Marysville, Michigan. This replacement would allow the company to achieve two goals: (1) Increase the amount of oil transported daily between the oil fields and the refineries. The replacement of the 24-inch-diameter pipe with a 36-inch pipe would enable the company to ship approximately twice the number of barrels per day (280,000 to 500,000 bpd) that it had previously moved. (An increase in diameter as small as two inches allows a pipeline to operate at higher pressure and move a lot more oil.) (2) Improve the integrity of the pipeline. The replacement of the forty-year-old pipeline with new pipe, using the latest materials and techniques, would make the pipeline less vulnerable to leaks and thereby safer.[31]

Enbridge had recognized the need to improve the integrity of the pipeline previously when a major spill occurred in July 2010 in Calhoun County, Michigan. Over 1 million gallons of oil leaked into the Kalamazoo River. Drinking water was contaminated, and homes had to be evacuated. The National Transportation Safety Board report indicated that the cause of the leak was metal corrosion and fatigue. The spill cost the company approximately $767 million, the costliest onshore spill in US history. Four years after the spill, the cleanup was still not completed.[32]

The Enbridge leak was one of an increasing number of pipeline leaks that have occurred recently. In May 2014 the Plains All American West Coast Pipeline—also known as Line 2000—spilled about 10,000 gallons into the streets of Los Angeles.[33] Residents over 1.5 miles away reported fumes and headaches associated with acute exposure to petrochemical toxins. Two months previously, in March 2014, between 7,000 and 10,000 gallons (26,000–38,000 liters) of sweet crude had leaked into the Oak Glen Nature Preserve in Southwestern Ohio about a quarter mile from the Great Miami River.[34]

When Enbridge began its plans to replace its pipe in the Indiana/Michigan area, it recognized that the communities around the pipeline would be hesitant about having a larger pipe after the damage done to the environment in Michigan as well as reports of the damage caused by other leaks. To reassure communities that the new pipeline, although carrying more oil, would be safer and greener than the old one, the company developed a presentation to explain the need for the expansion and to assure the various community groups involved of the pipeline's safety (see Figures 5.3–5.5).

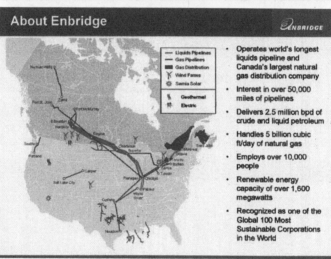

FIGURE 5.3A. *Chapter slide deck, slides 1–2*

Slide 1: Introductory slide contains three photographs to emphasize safety, cleanliness, and green space, the points to be emphasized in the presentation.

Slide 2: The slide provides an overview map, depicting the pipelines and other energy sources across the United States. This map serves as a framework for maps in the following slides. The text provides a key to the map. The text also provides secondary information in the right margin. The presenter uses a pointer to explain the areas of the map he is discussing and then each of the points listed in the right margin.

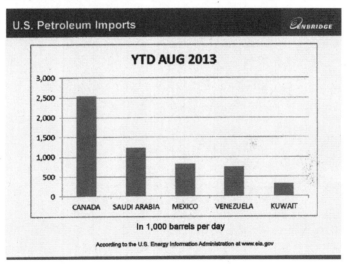

FIGURE 5.3B. *Chapter slide deck, slides 3–4*

Slide 3: The map is repeated but is smaller and has a different focus. Because the audience is already familiar with the map, it provides a frame of reference with which to learn the new information in the pullout.

Slide 4: The graph emphasizes the difference between the amount of oil produced in the United States and that in the remainder of the world.

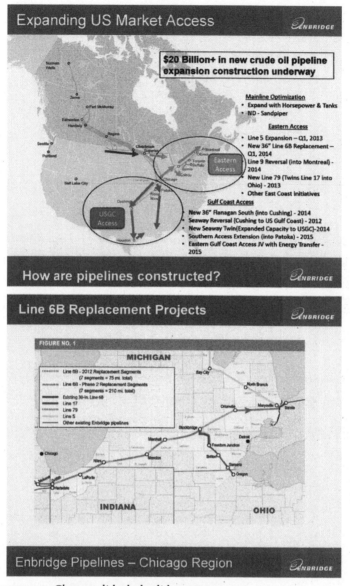

FIGURE 5.4A. *Chapter slide deck, slides 5–6*

Slide 5: *The same map is shown but with a different focus.*

Slide 6: *A new map of a section of the pipeline is shown but it is simply a slice of the previous map.*

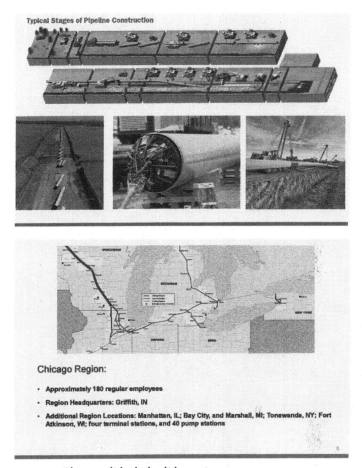

FIGURE 5.4B. *Chapter slide deck, slides 7–8*

Slide 7: Photographs and illustrations are used to show the parts of the pipeline.

Slide 8: Again, a slice of the map is shown here.

This was not mere spin. The spill had cost the company millions, and it was determined not only to make the pipeline as leakproof as possible but also to develop a system to contain a leak as quickly as possible if one did occur. It was these innovations that the company wanted to explain to its constituency.

The presentation provides an example of the effective integration of graphics and text in slideware. The visuals include photographs that

FIGURE 5.5A. *Chapter slide deck, slides 9–10*

Slide 9: The section on safety is given a separate cover page to emphasize its importance.

Slide 10: This photograph, providing an overview of the area, depicts a green landscape.

depict Enbridge's role in supplying energy to the nation, the construction of the new pipe, and the safety equipment to be installed as well as graphs that provide a visualization of statistical information. The text

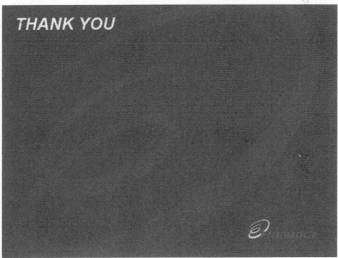

FIGURE 5.5B. *Chapter slide deck, slides 1–12*
Slide 11: This overview creates a picture of cleanliness and order.
Slide 12: This slide provides a courteous closing.

in the slides is minimal and is used mainly to provide captions for the graphics as well as the main points of the verbal presentation, which expands on the various points. Each slide focuses on a single major concept. The font for the headline for the section on safety appears as large as the title slide, emphasizing the point. It is this information that the company wants the audience to remember. Most of the slides also contain supplementary information that is not of major importance and is therefore placed in a secondary position (the right margin), often as a list of bullets. Photographs are used for the introductory slide and slides 9, 10, and 11 to emphasize the look of cleanliness and green space. The handouts are designed with four slides to a page, allowing space below each slide for note-taking.

Example of a Presentation with Text-Based Slideware

Using the guidelines for presenting text, this author might have developed an oral presentation based on this chapter with the slide deck and accompanying handouts in Figures 5.6a–5.6c.

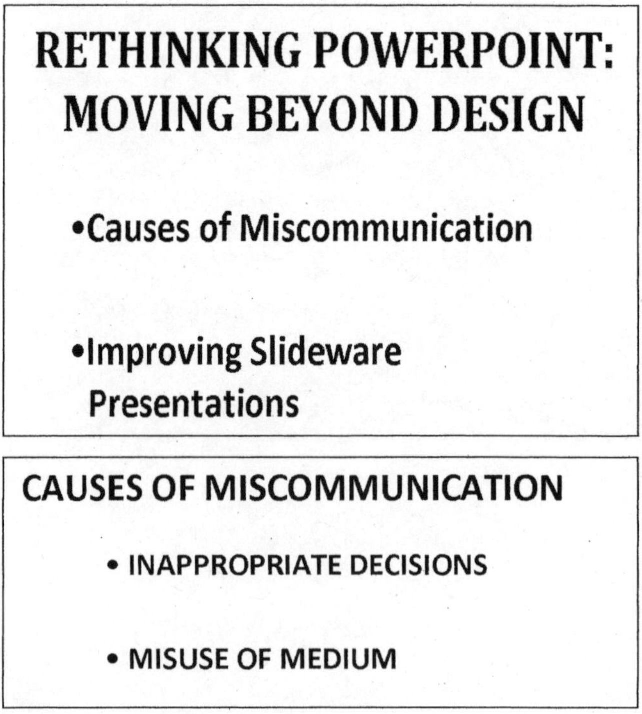

RETHINKING POWERPOINT: MOVING BEYOND DESIGN

- •Causes of Miscommunication

- •Improving Slideware Presentations

Slide 1

CAUSES OF MISCOMMUNICATION

- • INAPPROPRIATE DECISIONS

- • MISUSE OF MEDIUM

Slide 2

FIGURE 5.6A. *Chapter slide deck, slides 1–2*

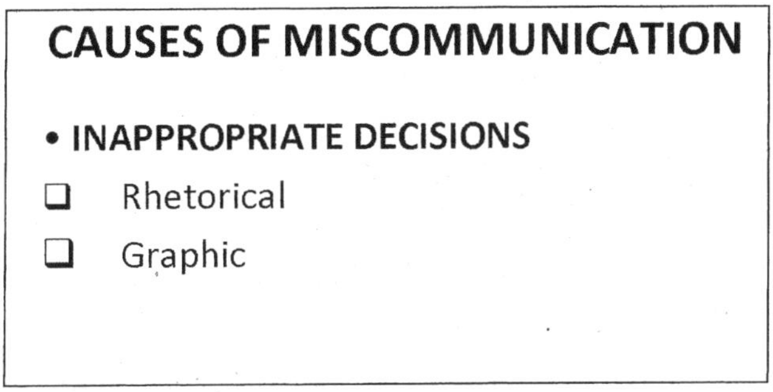
CAUSES OF MISCOMMUNICATION
• INAPPROPRIATE DECISIONS
Rhetorical
Graphic

Slide 3

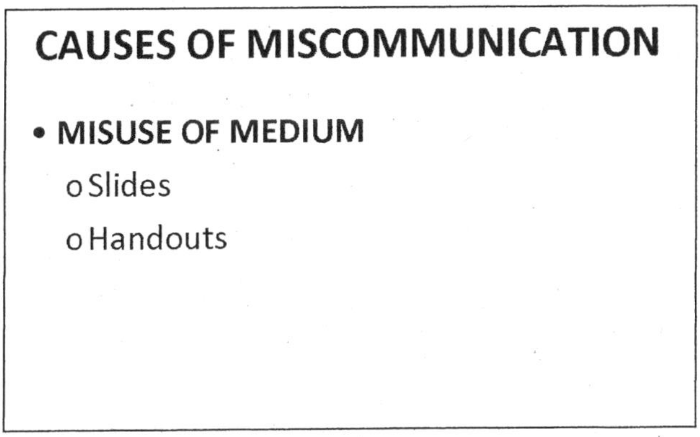
CAUSES OF MISCOMMUNICATION
• MISUSE OF MEDIUM
o Slides
o Handouts

Slide 4

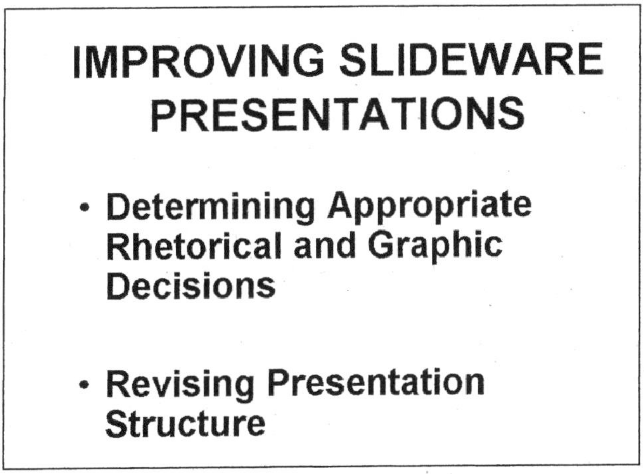
IMPROVING SLIDEWARE PRESENTATIONS
• Determining Appropriate Rhetorical and Graphic Decisions
• Revising Presentation Structure

Slide 5

FIGURE 5.6B. *Chapter slide deck, slides 3–5*

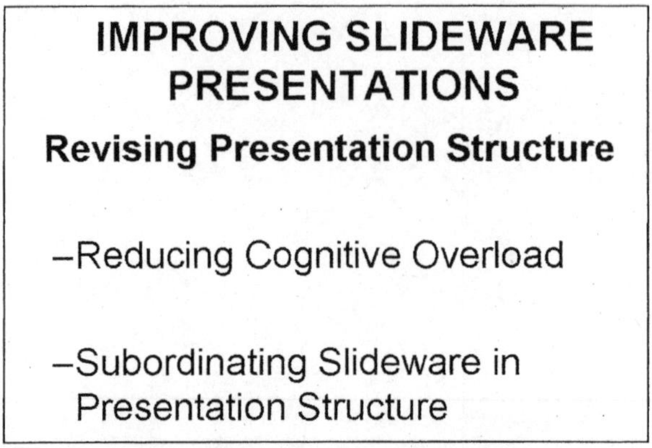

Slide 6

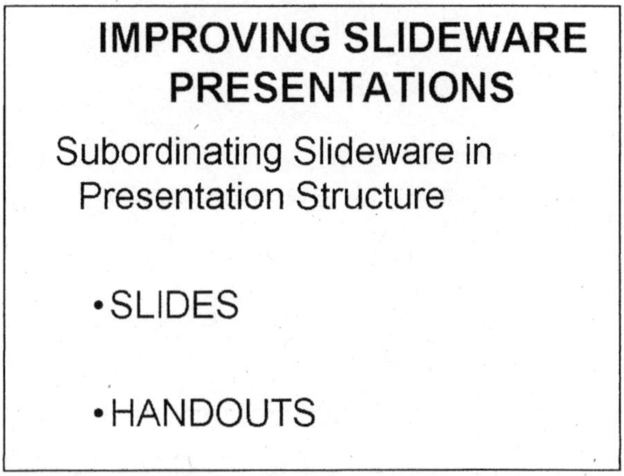

IMPROVING SLIDEWARE PRESENTATIONS
Revising Presentation Structure

–Reducing Cognitive Overload

–Subordinating Slideware in Presentation Structure

Slide 7

IMPROVING SLIDEWARE PRESENTATIONS

Subordinating Slideware in Presentation Structure

- SLIDES

- HANDOUTS

FIGURE 5.6C. *Chapter slide deck, slides 6–7*

Notes

1. Elisabeth Bumiller, "We Have Met the Enemy and He Is PowerPoint," *New York Times*, April 27, 2010, 1.

2. Marshall McLuhan, *Understanding Media: The Extensions of Man* (Boston: MIT Press, 1964).

3. I will use the term "reader/audience" to refer to the slides that an audience must read in addition to the oral presentation to which an audience must listen.

4. Columbia Accident Investigation Board, *Report of the Columbia Accident Investigation Board* (Washington, DC: Government Printing Office, 2003), 1.

5. Nancy Duarte, *Slide:ology: The Art and Science of Creating Great Presentations* (Sebastopol, CA: O'Reilly Media, 2008); Garr Reynolds, *Presentation Zen* (Berkeley, CA: New Riders, 2008).

6. Joanna K. Garner, Michael Alley, Allen F. Gaudelli, et al., "Common Use of PowerPoint versus the Assertion-Evidence Structure: A Cognitive Psychology Perspective," *Technical Communication* 56, no. 4 (November 2009): 331–45.

7. Jean-Luc Doumont, "The Cognitive Style of PowerPoint: Not All Slides Are Evil," *Technical Communication* (Washington) 52, no. 1 (February 2005): 64–70; Michael Markel, "Exploiting Verbal-Visual Synergy in Presentation Slides," *Technical Communication* (Washington) 56, no. 2 (May 2009): 122–31.

8. David K. Farkas, "A Brief Assessment of Michael Alley's Ideas Regarding the Design of PowerPoint Slides," 2010, http://faculty.washington.edu/farkas.

9. Edward Tufte, "PowerPoint Is Evil: Power Corrupts. PowerPoint Corrupts Absolutely," *Wired* 11, no. 9 (2003): 11–13, http://www.wired.com/wired/archive/11.09/ppt2_pr.html.

10. Columbia Accident Investigation Board, "Report of the Columbia Accident Investigation Board" (Washington, DC: Government Printing Office, 2003), 1.

11. Ibid., 143.

12. Ibid., 143–45.

13. Ibid., 191.

14. Markel, "Exploiting Verbal-Visual Synergy in Presentation Slides," 122–31.

15. Richard E. Mayer, ed., *The Cambridge Handbook of Multimedia Learning* (Cambridge: Cambridge University Press, 2005).

16. David K. Farkas, "Understanding and Using PowerPoint," *Society of Technical Communication*, 2005, http://faculty.washington.edu/farkas/Farkas-STC-05-UnderstandingPowerPoint.pdf; Farkas, "A Heuristic for Reasoning about PowerPoint Deck Design," *Proceedings of the 2008 Professional Communication Conference*, Montreal, July 13–16, 2008; Farkas, "Managing Three Mediation Effects That Influence PowerPoint Deck Authoring," *Technical Communication* (Washington) 56, no. 1 (February 2009): 28–38; Farkas, "A Brief Assessment of Michael Alley's Ideas Regarding the Design of PowerPoint Slides," 2010, http://faculty.washington.edu/farkas.

17. Garner et al., "Common Use of PowerPoint."

18. Allan Paivio, *Mental Representations* (New York: New York University Press, 1986).

19. John Sweller, "Cognitive Load Theory, Learning Difficulty, and Instructional Design," *Learning and Instruction* 4 (1994): 295–312.

20. Farkas, "Managing Three Mediation Effects."

21. Ian Kinchin, "Developing PowerPoint Handouts to Support Meaningful Learning," *British Journal of Educational Technology* 37 (2006): 647–50; Richard M. Felder and Rebecca Brent, "Death by PowerPoint: Tomorrow's Professor #689," https://tomprof.stanford.edu/posting/689, 2006; Jeffrey R. Young, "When Good Technology Means Bad Teaching," *The Chronicle of Higher Education*, 2004, http://chronicle.com/article /When-Good-Technology-Means-Bad/10922.

22. Michael J. Howe, "Using Students' Notes to Examine the Role of Individual Learning in Acquiring Meaningful Subject Matter," *Journal of Educational Research* 64, no. 2 (1970): 61–63.

23. Burke H. Bretzing and Raymond W. Kulhavy, "Notetaking and Depth of Processing," *Contemporary Educational Psychology* 4, no. 2 (1979): 145–53.

24. I. Phillip Ash and Bruce Carlton, "The Value of Note-Taking during Film Learning," *British Journal of Educational Psychology* 23, no. 2 (1953): 121–25.

25. I. Phillip Ash and Bruce Carlton, "Managing Three Mediation Effects That Influence PowerPoint Deck Authoring," *Technical Communication* (Washington) 56 (2009): 28–38.

26. Doumont, "The Cognitive Style of PowerPoint," 4–70.

27. Thomas Armstrong, *Multi-Intelligences in the Classroom*, 3rd ed. (Alexandria, VA: Association of Supervision and Curriculum Development, 2009).

28. Johnnie Scott, "The Topic Sentence Paragraph," accessed February 2016, http://www.csun.edu/~hcpas003/topic.html; Hussein Farhady and F. Sajadi, "Location of the Topic Sentence, Level of Language Proficiency, and Reading Comprehension," *Journal of the Faculty of Foreign Languages* (1999): 308–18, http://mpazhou.ir/wp-content/uploads/2011/11/Location -of-the-Topic-Sentence-Level-of-Language-Proficiency.pdf.

29. Illene Noppe, "PowerPoint Presentation Handouts and College Student Learning Outcomes," *International Journal for the Scholarship of Teaching and Learning*, 2007, http://www.georgiasouthern.edu; Jeff Beecher, *Notetaking: What Do We Know about the Benefits* (Bloomington, IN: ERIC Clearinghouse on Reading and Communication Skills, 1988), ERIC Identifier: ED300805, http://files.eric.ed.gov/fulltext/ED300805.pdf.

30. Avik Chowdhury, "Why Enbridge Is Expanding Pipeline Capacity in the Bakken," *Market Realist*, July 8, 2014, http://marketrealist.com/2014/07/why-enbridge-is-expanding-pipeline-capacity-in-bakken/.

31. Pipeline Safety Trust, Bellingham, WA, accessed February 2016, http://pstrust.org/about-pipelines1/enbridge-expansion-backgrounder; Joint Panel for the Enbridge Northern Gateway Project, "Report of the Joint Review Panel for the Enbridge Northern Gateway Project 2" (Calgary, Alberta: National Energy Board, 2013), http://gatewaypanel.review-examen.gc.ca/clf-nsi/dcmnt/rcmndtnsrprt/rcmndtnsrprt-eng.html.

32. National Transportation Safety Board, "Pipeline Accident Report: Enbridge Incorporated Hazardous Liquid Pipeline Rupture and Release," Marshall, MI, July 2012, http://www.ntsb.gov/investigations/Accident Reports/Pages/PAR1201.aspx; Steven Mufson, "NTSB Blames Enbridge 'Weak' Regulations in Kalamazoo Oil Spill," *Washington Post*, July 12, 2012.

33. Damon Nagami, "Oil Spill Threatens Los Angeles Neighborhoods and River Revitalization Efforts," May 15, 2014, http://switchboard.nrdc.org/blogs/dnagami/oil_spill_threatens_los_angele.html.

34. Dan Sewell, "Ohio Oil Spill: Mid-Valley Pipeline Leak Released More Than 20,000 Gallons into Oak Glen Preserve," Associated Press, May 24, 2014.

Bibliography

Achenbach, Joel. *A Hole at the Bottom of the Sea: The Race to Kill the BP Oil Gusher*. New York: Simon and Schuster, 2011.

Afflerbach, P. "The Influence of Prior Knowledge on Expert Readers' Main Idea Construction Strategies." *Reading Research Quarterly* 25, no. 1 (1990): 31–46. http://dx.doi.org/10.2307/747986.

Ambrose, Stephen. *Nothing Like It in the World: The Men Who Built the Transcontinental Railroad 1863–1869*. New York: Simon and Schuster, 2000.

American Public Power Association. "Electric Generating Utility (EGU) Mercury MACT Rule." February 2012. http://www.publicpower.org /files/PDFs/EGUMACTRuleFeb2012IB.pdf.

Armstrong, Thomas. *Multi-Intelligences in the Classroom*. 3rd ed. Alexandria, VA: Association of Supervision and Curriculum Development, 2009.

Ash, I. Phillip, and Bruce Carlton. "The Value of Note-Taking during Film Learning." *British Journal of Educational Psychology* 23, no. 2 (1953): 121–25. http://dx.doi.org/10.1111/j.2044-8279.1953.tb02848.x.

Associated Press. "Gulf Oil Spill Deaths: The 11 Rig Workers Who Died during the BP Deepwater Horizon Explosion." *Huffington Post*, November 15, 2012. www.huffingtonpost.com/2012/11/15/gulf-oil-spill-deaths _n_2139669.html.

Barley, Stephen R., Deborah E. Myerson, and Stine Grodal. "Email as a Source and Symbol of Stress." August 2011. http://people.bu.edu/grodal /Email.pdf.

DOI: 10.5876/9781607324676.c006

Bartlit, Beck, Palenchar, Herman & Scott, LLP. "Presidential Oil Spill Commission Report from Chief Council, Fred Bartlit." February 2011.

Beecher, Jeff. "Notetaking: What Do We Know about the Benefits." Bloomington, IN: ERIC Clearinghouse on Reading and Communication Skills, 1988. ERIC Identifier: ED300805. http://files.eric.ed.gov/fulltext/ED300805.pdf.

Bipartisan Policy Center. "Assessment of EPA's Utility MACT Proposal." March 2011. http://bipartisanpolicy.org/wp-content/uploads/sites/default/files/Q&A%20Assessment%20of%20MACT%20Rule.pdf.

Boiarsky, C. *Technical Writing: Contexts, Audiences and Communities.* Boston: Allyn and Bacon, 1993.

Bretzing, Burke H., and Raymond W. Kulhavy. "Notetaking and Depth of Processing." *Contemporary Educational Psychology* 4, no. 2 (April 1979): 145–53.

Bronstein, Scott, and Wayne Drash. "Rig Survivors: BP Ordered Shortcut on Day of Blast." *CNN*, June 9, 2011. http://www.cnn.com/2010/US/06/08/oil.rig.warning.signs/.

Brown, Carolyn M., and Robin B. Thomerson. "United States Supreme Court Reverses Utility MACT Rule." *National Law Review*, June 29, 2015. http://www3.epa.gov/mats/actions.html.

Bryce, Robert. "Dirty but Essential—That's Coal." *Los Angeles Times*, July 27, 2012. http://articles.latimes.com/2012/jul/27/opinion/la-oe-adv-bryce-coal-epa-climate-20120727.

Bullett, Karla. "US Army Corps of Engineers Responds to Strange Flood," Chicago District, archived from the original on June 7, 2008. May 2002.

Bumiller, Elisabeth. "We Have Met the Enemy and He is PowerPoint." *New York Times*, April 27, 2010, 1.

Byron, Kristin. "Carrying Too Heavy a Load? The Communication and Miscommunication of Emotion by Email." *Academy of Management Review* 33, no. 2 (2008): 309–27. http://dx.doi.org/10.5465/AMR.2008.31193163.

Carlson, Douglas, Marty Horn, Thomas Van Biersel, and David Fruge. "Atchafalaya Basin Inundation Data Collection and Damage Assessment Project." Baton Rouge, LA: Louisiana Geological Survey, 2011. http://data.dnr.louisiana.gov/ABP-GIS/ABPstatusreport/Report_of_Investigation_12-01web.pdf.

Carr, Nicholas. *The Shallows: What the Internet Is Doing to Our Brains.* New York: W. W. Norton and Company, 2011.

Chalmers, Stuart. "Dang, I Hit the Send Button Too Soon." Accessed February 2016. http://digitalbloggers.com/normalguy/dang-i-hit-the-send-button-too-soon/.

Chaparro, B. S., A. D. Shaikh, and A. Chaparro. "Examining the Legibility of Two New ClearType Fonts." February 2006. http://usabilitynews.org /examining-the-legibility-of-two-new-cleartype-fonts/.

Chowdhury, Avik. "Why Enbridge Is Expanding Pipeline Capacity in the Bakken." *Market Realist*, July 8, 2014. http://marketrealist.com/2014/07 /why-enbridge-is-expanding-pipeline-capacity-in-bakken/.

Columbia Accident Investigation Board. Report of the Columbia Accident Investigation Board. Vol. 1. Washington, DC: Government Printing Office, 2003.

Coney, Mary B. "Technical Readers and Their Rhetorical Roles." *IEEE Transactions on Professional Communication* 35, no. 2 (June 2, 1992): 58–63. http://dx.doi.org/10.1109/47.144864.

Conniff, Richard. "The Myth of Clean Coal." *Yale Environment 360*, June 3, 2008. http://e360.yale.edu/feature/the_myth_of_clean_coal/2014/.

Dabbish, Laura A., Robert E. Kraut, Susan Fussell, and Sara Kiesler. "Understanding Email Use: Predicting Action on a Message." 2005. http://www .cs.cmu.edu/~kiesler/publications/2005pdfs/2005-Dabbish-CHI.pdf. http://dx.doi.org/10.1145/1054972.1055068.

Daley, Richard. "Statement of Mayor Richard M. Daley, Preliminary Inquiry Update P/C." April 22, 1992.

Daley, Richard M. "Statement of Mayor Richard Daley, Preliminary Investigation on Flooding P/C." April 14, 1992.

Department of Energy. "The Clean Coal Technology Program." February 12, 2013. http://www.fe.doe.gov/education/energylessons/coal/coal_cct2 .html.

Derks, Daantje, and Arnold B. Bakker. "The Impact of E-mail Communication on Organizational Life." *Cyberpsychology (Brno)*, May 2013. http:// www.cyberpsychology.eu/view.php?cisloclanku=2010052401 &article=4.

Dolkart, Leo. "The Old Chicago Tunnel." *Midwest Engineer* 7 (December 1963): 22.

Dombrowski, Paul. "The Lessons of the Challenger Investigations." *Professional Communication IEEE Transactions* 34, no. 4 (1991): 211–16. http:// dx.doi.org/10.1109/47.108666.

Doumont, Jean-Luc. "The Cognitive Style of PowerPoint: Not All Slides Are Evil." *Technical Communication* (Washington) 52, no. 1 (February 2005): 64–70.

Duarte, Nancy. *Slide:ology: The Art and Science of Creating Great Presentations.* Sebastopol, CA: O'Reilly Media, 2008.

Ekroth, Lauren. "Have Email Conversation Problems?" *DonMorris.* http:// donmorris.com/article/have-email-conversation-problems.

Estes, Janelle. "Email Subject Lines: 5 Tips to Attract Readers." *eNewsline*. May 4, 2014. http://www.nngroup.com/articles/email-subject-lines/.

Farhady, Hussein, and F. Sajadi. "Location of the Topic Sentence, Level of Language Proficiency, and Reading Comprehension," 308–18, 1999. http://mpazhou.ir/wp-content/uploads/2011/11/Location-of-the-Topic-Sentence-Level-of-Language-Proficiency.pdf.

Farkas, David K. "A Brief Assessment of Michael Alley's Ideas Regarding the Design of PowerPoint Slides." 2010. http://faculty.washington.edu/farkas.

Farkas, David K. "A Heuristic for Reasoning about PowerPoint Deck Design." *Proceedings of the 2008 Professional Communication Conference*, Montreal, Canada, July 13–16, 2008.

Farkas, David K. "Managing Three Mediation Effects That Influence Power-Point Deck Authoring." *Technical Communication* (Washington) 56, no. 1 (February 2009): 28–38.

Farkas, David K. "Understanding and Using PowerPoint." *Society of Technical Communication*, 2005. http://faculty.washington.edu/farkas/Farkas-STC-05-UnderstandingPowerPoint.pdf.

Fears, Darryl. "Deepwater Horizon Oil Left Tuna, Other Species with Heart Defects Likely to Prove Fatal." *Washington Post*, March 24, 2014. http://www.washingtonpost.com/national/health-science/after-deepwater-oil-spill-once-speedy-tuna-no-longer-make-the-grade/2014/03/24/4d2e2d78-b378-11e3-b899-20667de76985_story.html.

Felder, Richard M., and Rebecca Brent. "Death by PowerPoint." *Tomorrow's Professor #689*, 2006. Tomorrows-professor@lists.Stanford.edu.

Flower, Linda. "Writer-Based Prose: A Cognitive Basis for Problems in Writing." *College English* 41, no. 1 (1979): 19–37. http://dx.doi.org/10.2307/376357.

Garner, Joanna K., Michael Alley, Allen F. Gaudelli, et al. "Common Use of PowerPoint versus the Assertion-Evidence Structure: A Cognitive Psychology Perspective." *Technical Communication* 56, no. 4 (November 2009): 331–45.

Geman, Ben, and Nathaniel Gronewold. "Coal Fired Power Plants Will Need Better Carbon Capture and Storage Technology." *Scientific American*, February 12, 2009. http://www.scientificamerican.com/article/coal-fired-power-plants-carbon-capture/.

Ghosh, Tanu, JoAnne Yates, and Wanda Orlikowski. "Using Communication Norms for Coordination: Evidence from a Distributed Team." *Research Brief*. Boston, MA: MIT Center for eBusiness, October 2004, http://aisel.aisnet.org/icis2004/10.

Gócza, Zoltán. "Myth #3: People Don't Scroll." *UX Myths*, 2012. http://uxmyths.com/post/654047943/myth-people-dont-scroll.

Hall, Edward. *Beyond Culture*. New York: Anchor Books, 1976.

Halliday, M.A.K., and Ruqaiya Hasan. *Cohesion in English*. New York: Pearson Education Ltd., 1976.

Hastings, Sally O., and Holly J. Payne. "Expressions of Dissent in Email: Qualitative Insights into Uses and Meanings of Organizational Dissent." *Journal of Business Communication* 50, no. 3 (2013): 309–31. http://dx.doi.org/10.1177/0021943613487071.

Herndl, Carl, Barbara A. Fennell, and Carolyn R. Miller. "Understanding Failures in Organizational Discourse: The Accident at Three Mile Island and the Shuttle Challenger Disaster." In *Text and the Professions*, ed. C. Bazerman and J. Paradis, 279–305. Madison: University of Wisconsin Press, 1991.

Hiltzik, Michael. *Colossus: Hoover Dam and the Making of the American Century*. New York: Free Press, 2010.

Hopkins, Andrew. *Disastrous Decisions: The Human and Organisational Causes of the Gulf of Mexico Blowout*. Sydney: CCH Australia Ltd., 2012.

Hopkins, Andrew. *Failure to Learn: The BP Texas City Refinery Disaster*. Sydney: CCH Australia Ltd, 2010.

Howe, Michael J. "Using Students' Notes to Examine the Role of Individual Learning in Acquiring Meaningful Subject Matter." *Journal of Educational Research* 64, no. 2 (1970): 61–63. http://dx.doi.org/10.1080/00220671.1970.10884094.

Huckin, Thomas N. "A Cognitive Approach to Readability." In *New Essays in Technical and Scientific Communication: Research, Theory, and Practice*, ed. Paul V. Anderson, R. John Brockman, and Carolyn R. Miller. Farmingdale, NY: Baywood, 1983.

Inouye, Randall R., and Joseph D. Jacobazzi. "The Great Chicago Flood of 1992." *Civil Engineering–ASCE* 62, no. 11 (November 1992): 52–55.

Institute for Energy Research. "Coal." *Encyclopedia*. Washington, DC, 2013. http://instituteforenergyresearch.org/topics/encyclopedia/coal/.

Jabr, Ferris. "The Reading Brain in the Digital Age: The Science of Paper versus Screens." *Scientific American*, April 11, 2013. http://www.scientificamerican.com/article/reading-paper-screens/.

Jackson, David. "In City Hall Memos, Everything Is 'Serious.'" *Chicago Tribune*, April 28, 1992. http://articles.chicagotribune.com/1992-04-26/news/9202070124_1_memo-city-hall-flood.

Jackson, Thomas. "Email Stress." 2012. http://www.profjackson.com/email_stress.html.

Joint Panel for the Enbridge Northern Gateway Project. "Report of the Joint Review Panel for the Enbridge Northern Gateway Project 2." Calgary,

Alberta: National Energy Board, 2013. http://gatewaypanel.review
-examen.gc.ca/clf-nsi/dcmnt/rcmndtnsrprt/rcmndtnsrprt-eng.html.

Kahan, Dan. "Fixing the Communications Failure." *Nature* 463, no. 7279
(2010): 296–7. http://dx.doi.org/10.1038/463296a.

Kahan, Dan, and Donald Braman. "Cultural Cognition and Public Policy."
Yale Law & Policy Review 24, no. 47 (2006): 148–70.

Keohane, Joe. "How Facts Backfire." *Boston.com*, July 11, 2010. http://www
.boston.com/bostonglobe/ideas/articles/2010/07/11/how_facts_backfire/.

Kinchin, Ian. "Developing PowerPoint Handouts to Support Meaningful
Learning." *British Journal of Educational Technology* 37, no. 4 (2006):
647–50. http://dx.doi.org/10.1111/j.1467-8535.2006.00536.x.

Kleiner, Perkins, Caulfield, and Byers. "KPCB Internet Trends." May 29, 2013.
http://www.slideshare.net/kleinerperkins/kpcb-internet-trends-2013.

Konrad, John, and Tom Shroder. *Fire on the Horizon: The Untold Story of the
Gulf Oil Disaster*. New York: HarperCollins Publishers, 2011.

Kooti, Farshad, Luca Maria Aiello, Mihajito Grbovic, et al. "Evolution of
Conversations in the Age of Email Overload." Proceedings of the 24th
International Conference World Wide Web. Florence, Italy (2015):
603–13. http://www-scf.usc.edu/~kooti/files/kooti_email.pdf.

Kunzelman, Michael. "Rig Supervisors Robert Kaluza and Donald Vidrine
Challenge Manslaughter Counts." *Huffington Post*, May 13, 2013. http://
www.huffingtonpost.com/2013/05/31/bp-rig-supervisors-robert-kaluza
-donald-vidrine_n_3366355.html.

Kupritz, Virginia. "Productive Management Communication: Online and
Face-to-Face." *International Journal of Business Communication* 48, no. 1
(January 2011): 54–82. http://job.sagepub.com/content/48/1/54.abstract.

Langer, J. "The Reading Process." In *Secondary School Reading: What
Research Reveals about Classroom Practice*, ed. A. Berger and H. A. Robinson, 39–52. Urbana, IL: National Council of Teachers of English, 1982.

Louhiala-Salminen, L. "From Business Correspondence to Message
Exchange: What Is There Left?" In *Business English: Research into Practice*,
ed. C. Nickerson and M. Hewings, 100–114. New York: Longman, 1999.

Markel, Michael. "Exploiting Verbal-Visual Synergy in Presentation Slides."
Technical Communication (Washington) 56, no. 2 (May 2009): 122–31.

Marketing Charts Staff. "When Smartphone Users Check Email during the
Day." *American Writers and Artists Inc.* May 2014. http://www.marketing
charts.com/online/when-smartphone-users-check-email-during-the-day
-41401/.

Mayer, Richard E., ed. *The Cambridge Handbook of Multimedia Learning*.
Cambridge: Cambridge University Press, 2005. http://dx.doi.org/10.1017
/CBO9780511816819.

McDermott, Will, and Emery McDermott. Letter from McDermott, Will and Emery, to Janet M. Koran, April 20, 1992. Outline of events (FOIA [Freedom of Information Act] 003444).

McLuhan, Marshall. *Understanding Media: The Extensions of Man*. Boston: MIT Press, 1964.

McMurrey, David. "Audience Analysis: Just Who Are These Guys?" 2015. https://learn.saylor.org/mod/page/view.php?id=5540.

McRaney, David. "The Backfire Effect." *You Are Not So Smart*, June 10, 2011. http://youarenotsosmart.com/2011/06/10/the-backfire-effect/.

Melcrum. "Choosing the Right Communication Channel." 2015. https://www.melcrum.com/research/strategy-planning-tactics-intranets-digital-social-media/choosing-right-communication.

Miss Manners. "Stop, Reread and Think before You Hit Send." 2012. http://www.businessemailetiquette.com/stop-reread-and-think-before-you-send/2012.

Moffatt, Bruce G. "The Chicago Freight Tunnels." Spring 2011. http://www.mascontext.com/issues/9-network-spring-11/the-chicago-freight-tunnels/.

Mufson, Steven. "NTSB Blames Enbridge 'Weak' Regulations in Kalamazoo Oil Spill." *Washington Post*, July 12, 2012.

Mujumdar, Anujeet. "Smartphone Users Check Their Phones an Average of 150 Times a Day." *Tech2*, May 30, 2013. http://tech.firstpost.com/news-analysis/smartphone-users-check-their-pnones-an-average-of-150-times-a-day-86984.html.

Nagami, Damon. "Oil Spill Threatens Los Angeles Neighborhoods and River Revitalization Efforts." May 15, 2014. http://switchboard.nrdc.org/blogs/dnagami/oil_spill_threatens_los_angeles.html.

National Academy of Sciences. *Proceedings of the National Academy of Sciences* 111, no. 32 (2014): 128, 193–194, 199–202.

National Commission on the BP Deepwater Horizon Oil Spill and Offshore Drilling. *Deepwater: The Gulf Oil Disaster and the Future of Offshore Drilling; Report to the President*. Washington, DC: Government Printing Office, 2011.

National Dissemination Center for Children with Disabilities. "How People Read on the Web." *Center for Parent Information and Resources*. August 2012. http://www.parentcenterhub.org/repository/web-reading.

National Mining Association. "Clean Coal Technology." February 12, 2013. Washington, DC: Department of Energy, 2013. http://www.nma.org/pdf/fact_sheets/cct.pdf.

National Oceanic and Atmospheric Administration. "United States Flood Loss Report—Water Year 2011." 2013. http://www.nws.noaa.gov/hic /summaries/WY2011.pdf.

National Transportation Safety Board. "Pipeline Accident Report: Enbridge Incorporated Hazardous Liquid Pipeline Rupture and Release," Marshall, MI, July 2012. http://www.ntsb.gov/investigations/AccidentReports /Pages/PAR1201.aspx.

Nielsen, Jakob. "F-Shaped Pattern for Reading Web Content." *Nielsen Norman Group*, April 2006. http://www.nngroup.com/articles/f-shaped -pattern-reading-web-content/.

Nielsen, Jakob. "How Users Read on the Web." *Nielsen Norman Group*, October 1997. http://www.nngroup.com/articles/how-users-read-on-the-web/.

Noppe, Illene. "PowerPoint Presentation Handouts and College Student Learning Outcomes." *International Journal for the Scholarship of Teaching and Learning.* 2007. http://www.georgiasouthern.edu.

Nyhan, Brendan, Jason Reifler, Sean Richey, and Gary Freed. "Effective Messages in Vaccine Promotion: A Randomized Trial." *Pediatrics.* Elk Grove Village, IL: Journal of the American Academy of Pediatrics, 2014. http:// pediatrics.aappublications.org/content/early/2014/02/25/peds.2013 -2365.

O'Brien, Ellen, and Lyle Benedict. "1992, April 13: Freight Tunnel Flood." Chicago Public Library, 2005. http://www.chipublib.org/004chicago /disasters/tunnel_flood.html.

Outing, Steve. "Eyetrack III: What News Websites Look Like through Readers' Eyes." *Poynter*, March 2011. http://www.poynter.org/uncatego rized/24963/eyetrack-iii-what-news-websites-look-like-through-readers -eyes/.

Paivio, Allan. *Mental Representations*. New York: New York University Press, 1986.

Pallardy, Richard. "Mississippi Flood of 2011." *Encyclopedia Britannica*, October 2013. http://www.britannica.com/event/Mississippi-River-flood-of-2011.

Pearson, David. "Readers and Contexts of Use." http://www.pearsonhigher ed.com/samplechapter/0205632440_ch3.pdf.

Pipeline Safety Trust. Bellingham, WA. Accessed February 2016. http:// pstrust.org/about-pipelines1/enbridge-expansion-backgrounder.

Presidential Commission on the Space Shuttle Challenger Disaster. *The Report of the Presidential Commission on the Space Shuttle Challenger Accident*. Washington, DC: Government Printing Office, 1986.

President's Commission on the Accident at Three Mile Island. "Report of the President's Commission on the Accident at Three Mile Island." 1979. http://www.threemileisland.org/downloads/188.pdf.

Quinn, Sara Dickenson. "New Poynter Eyetrack Research Reveals How People Read News on Tablets." *Poynter*, October 2012. http://www.poynter.org/how-tos/newsgathering-storytelling/visual-voice/191875/new-poynter-eyetrack-research-reveals-how-people-read-news-on-tablets/.

Radicati Group. "Email Statistics Report, 2011–2015." May 2011. http://www.radicati.com/wp/wp-content/uploads/2011/05/Email-Statistics-Report-2011-2015-Executive-Summary.pdf.

Reynolds, Garr. *Presentation Zen*. Berkeley, CA: New Riders, 2008.

Rioux, Paul. "Morganza Floodway Opens to Divert Mississippi River Away from Baton Rouge, New Orleans." *Times Picayune*, May 14, 2011. http://www.nola.com/environment/index.ssf/2011/05/morganza_floodway_opens_to_div.html.

Ripley, Amanda. *The Unthinkable: Who Survives When Disaster Strikes—and Why*. New York: Three Rivers Press, 2009.

Rogers, Hal. "Rogers Statement on EPA Issuance of Utility MACT Rule." December 21, 2011. http://halrogers.house.gov/news/documentsingle.aspx?DocumentID=273432.

Roschelle, Jeremy. "Learning in Interactive Environments: Prior Knowledge and New Experience." 1995. https://www.sri.com/sites/default/files/publications/imports/RoschellePriorKnowledge.pdf.

Rowan, Katherine E. "Earning Trust and Productive Partnering with the Media and Public." *Consortium of Social Science Associations*, 2004. http://www.cossa.org/seminarseries/risk_and_crisis.htm.

Ryan, Margaret. "EPA MACT Rule Released: Coal Plants Set for Closure as Blackout Risks Cited." *Breaking Energy*, December 21, 2011. http://breakingenergy.com/2011/12/21/epa-utility-mact-rule-released-coal-plants-set-for-closure-as-b/.

Sandman, Peter M. *Responding to Community Outrage: Strategies for Effective Risk Communication*. Fairfax, VA: American Industrial Hygiene Association, 1993. http://dx.doi.org/10.3320/978-0-932627-51-3.

Scerri, Simon. *Aiding the Workflow of Email Conversations by Enhancing Email with Semantics*. Galway, Ireland: Digital Enterprise Research Institute, 2007. http://aran.library.nuigalway.ie/xmlui/handle/10379/500.

Schleifstein, Mark. "Mississippi Flooding in New Orleans Area Could Be Massive If Morganza Floodway Stays Closed." *Times Picayune*, 2011. www.nola.com/weather/index.ssf/2011/05/army_corps_fears_massive_flood.html.

Scott, Johnnie. "The Topic Sentence Paragraph." Accessed February 2016. http://www.csun.edu/~hcpas003/topic.html.

Scribner, Sylvia, and Michael Cole. *The Psychology of Literacy*. Cambridge, MA: Harvard University Press, 1981. http://dx.doi.org/10.4159/harvard.9780674433014.

Sewell, Dan. "Ohio Oil Spill: Mid-Valley Pipeline Leak Released More Than 20,000 Gallons into Oak Glen Preserve." Associated Press, May 24, 2014.

Shaikh, A. Dawn. "The Effects of Line Length on Reading Online News." July 2005. http://psychology.wichita.edu/surl/usabilitynews/72 /linelength.asp.

Silverman, David, and William Gaines. "Flood Just a Matter of Inches." *Chicago Tribune*, April 26, 1992, 1.

Skovholt, Karianne. "Email Literacy in the Workplace." Dissertation, University of Oslo, 1981.

Sneed, Michael. "The River's Edge." *Chicago Sun Times*, April 16, 1992.

Sourcewatch. "American Coalition for Clean Coal Electricity." July 2012. http://www.sourcewatch.org/index.php?title=American_Coalition _for_Clean_Coal_Electricity.

Sourcewatch. "Environmental Impacts of Coal." March 2015. http://www .sourcewatch.org/index.php/Environmental_impacts_of_coal.

Stack, Laura. "Choose the Most Productive Communication Channel for Your Message." http://office.microsoft.com/en-us/help/choose-the-most -productive-communication-channel-for-your-message-HA001194524 .aspx (no longer available).

Sweller, John. "Cognitive Load Theory, Learning Difficulty, and Instructional Design." *Learning and Instruction* 4 (1994): 295–312.

Tedford, Deborah. "Why We Still Mine Coal." Washington, DC: National Public Radio. August 8, 2010. http://www.npr.org/templates/story/story .php?storyId=125694190.

TOTSE. "The Great Chicago Flood of 1992." http://totse.mattfast1.com/en /politics/political_spew/chiflood.html.

Townson, Patrick. The Great Chicago Flood of 1992. 1992. http://totse .mattfast1.com/en/politics/political_spew/chiflood.html.

Tufte, Edward. "PowerPoint Is Evil: Power Corrupts. PowerPoint Corrupts Absolutely." *Wired* 11, no. 9 (2003): 11–13. http://www.wired.com/wired /archive/11.09/ppt2_pr.html.

Turnage, Anna K., and Alan K. Goodboy. "E-Mail and Face-to-Face Organizational Dissent as a Function of Leader-Member Exchange Status." April 2014. http://job.sagepub.com/content/early/2014/04/13/2329488414525 456.full.pdf. http://dx.doi.org/10.1177/2329488414525456.

Union of Concerned Scientists. "Coal Is a Dirty Energy Source." Cambridge, MA. January 2014. http://www.ucsusa.org/clean_energy/smart-energy -solutions/decrease-coal/.

University of Bath. "Effective Email." 2016. http://www.bath.ac.uk/bucs /email/guidelines/effectiveemail.

US Army Corps of Engineers. "Morganza Floodway Proposed Clarifications to Standing Instructions." 2011. http://www.mvn.usace.army.mil/Missions /MississippiRiverFloodControl/MorganzaFloodwayOverview.aspx.

Wald, Matthew L. "G.M. Illustrates Managers' Disconnect." *Chicago Tribune*, June 9, 2014, B3. http://www.nytimes.com/2014/06/09/business /gm-report-illustrates-managers-disconnect.html.

Wood, John. "The Best Fonts to Use in Print, Online and Email." *American Writers and Artists Inc.* October 2011. http://www.awaionline.com /2011/10/the-best-fonts-to-use-in-print-online-and-email/.

Young, Jeffrey R. "When Good Technology Means Bad Teaching." *The Chronicle of Higher Education*, 2004. http://chronicle.com/article/When -Good-Technology-Means-Bad/10922.

Websites with Background and Additional Information

These websites provide videos, news stories, and updates on the topics discussed in this book. You can also find background and additional information on each topic as well as read the full reports of events that are discussed.

Introduction

Daiichi, Japan, nuclear accident: http://spectrum.ieee.org/energy/nuclear /24-hours-at-fukushima

Three Mile Island, Pennsylvania, nuclear accident: http://www.threemile island.org/downloads/188.pdf

Massey, West Virginia, coal mine explosion: http://www.msha.gov/Fatals /2010/UBB/FTL10c0331noappx.pdf

British Petroleum (BP) Gulf oil spill: http://www.gpo.gov/fdsys/pkg/GPO -OILCOMMISSION/pdf/GPO-OILCOMMISSION.pdf

Protest over megadairy project in northwest Illinois: http://www .stopthemegadairy.org/

Chapter 1: Writing and Reading in the Context of the Environmental Sciences

Lower Mississippi River flood of 2011: http://www.britannica.com/event /Mississippi-River-flood-of-2011

GM report illustrates managers' disconnect: http://www.nytimes.com/2014 /06/09/business/gm-report-illustrates-managers-disconnect.html?_r=0

Great Chicago Flood of April 13, 1992: http://www.worldcat.org/title/great -chicago-flood/oclc/316804963

Chicago freight tunnels: http://www.mascontext.com/issues/9-network -spring-11/the-chicago-freight-tunnels/

Memos an art form at city hall: http://articles.chicagotribune.com/1992-04 -26/news/9202070124_1_memo-city-hall-flood

President's Commission on the Three Mile Island Nuclear Accident report for the president: http://www.threemileisland.org/downloads/188.pdf

Chapter 2: Effective Discourse Strategies

Fight against fracking: http://www.thenation.com/article/fight-against -fracking/

Facts on the hydraulic fracturing process: http://www.exxonmobilperspectives .com/2011/06/17/facts-hydraulic-fracturing-process/?gclid=CO3h7O2Zx b8CFSJo7Aodcj8Aqw&gclsrc=aw.ds

Scientists warn of risk from fracking operations: http://news.national geographic.com/news/energy/2014/05/140502-scientists-warn-of-quake -risk-from-fracking-operations/

How oil and gas disposal wells can cause earthquakes: http://stateimpact.npr .org/texas/tag/earthquake/

The neglected human costs of transporting oil and gas: www.hhrjournal.org /2014/07/01/wrong-side-of-the-tracks-the-neglected-human-costs-of -transporting-oil-and-gas-2/

Future flood losses in major coastal cities: http://www.nature.com/nclimate /journal/v3/n9/full/nclimate1979.html

2011 Mississippi Flood: http://www.nws.noaa.gov/hic/summaries/WY2011 .pdf

Morganza Floodway proposed clarifications to standing instructions: http:// www.mvn.usace.army.mil/Missions/MississippiRiverFloodControl /MorganzaFloodwayOverview.aspx

Rivers to Ridges program: http://issuu.com/msaprofessionalservices/docs
/rivers2ridges_final_may_2008

Coherence and cohesion in text linguistics: http://www.criticism.com/da
/coherence.php

Eight kernels from the psychology of language: http://books.google.com
/books?id=Ip5cAgAAQBAJ&pg=PA40&lpg=PA40&dq=Define+
sentence+kernel+chomsky&source=bl&ots=orinG23w4I&sig=paq
PKGqBT3ZRaeI7bd5dkYkPsVk&hl=en&sa=X&ei=wxbMU7HtA-zJsQ
TMnIJo&ved=oCFgQ6AEwBw#v=onepage&q&f=false

Chapter 3: Effective Persuasive Strategies

Coal facts: http://www.c2es.org/energy/source/coal

American Coalition for Clean Coal Electricity (ACCCE): http://americas
power.org/

Judge halts killing wolves to save Alaskan caribou: http://blogs.findlaw.com
/decided/2010/06/judge-halts-killing-wolves-to-save-alaskan-caribou
.html

Energy Cost Impacts on American Families: http://americaspower.org/sites
/default/files/Energy_Cost_Impacts_2012_FINAL.pdf

National Economic Research Associates (NERA): http://www.nera.com
/index.htm

Modern Language Association (MLA) conventions: https://owl.english
.purdue.edu/owl/resource/747/01/

American Society of Mechanical Engineers (ASME) author guidelines:
https://www.asme.org/shop/proceedings/conference-publications
/author-guidelines

Institute of Electrical and Electronic Engineers (IEEE) author information:
http://www.ieee.org/publications_standards/publications/authors/index
.html

National Association of Environmental Professionals (NAEP) submission
guidelines: http://www.naep.org/assets/documents/guidelinesforEP
journalsubmissions-2.1.10.pdf

American Academy of Environmental Engineers and Scientists (AAEES):
http://www.aaees.org/

American Society of Plant Biologists (ASPB) author guidelines: http://www
.bioone.org/page/arbo.j/authors

Chapter 4: Communicating with Electronic Media

BP/*Horizon* Gulf oil rig explosion: http://www.gpo.gov/fdsys/pkg
/GPO-OILCOMMISSION/pdf/GPO-OILCOMMISSION.pdf
Columbia shuttle accident: http://www.nasa.gov/columbia/home/CAIB
_Vol1.html
Choosing a communication channel: flowchart: http://www.covenanthealth
.org/Covenant-University/Leadership-Learning-Center/Professional
-Development/Communication-Channel.aspx

Chapter 5: Communicating with PowerPoint

Presentation to senior military officers analyzing the dynamics and chal-
lenges of the counterinsurgency (COIN) in Afghanistan: http://www.oss
.net/dynamaster/file_archive/091222/9649ad18a0d538dc213e13af676e3aa8
/Afghanistan_Dynamic_Planning.pdf
Don McMillan's "Life after Death by PowerPoint": https://www.youtube
.com/watch?v=lpvgfmEU2Ck
Report of *Columbia* Accident Investigation Board: http://www.nasa.gov
/columbia/home/CAIB_Vol1.html
Expansion of the Enbridge Pipeline: http://www.enbridge.com/
Replacement of the forty-year-old pipeline: http://www.enbridge.com
/Line6BReplacementProjects.aspx

Index